Living Water

A Holistic Perspective of Microeconomics and Macroeconomics

Cover Theme: New Life via Living Water Baptism

Global economy is plagued with human greed, distorted value system and misleading theories. The cover artwork urges the world to take the step of exodus to transform from old to new via the *living water baptism*. This book provides a holistic perspective of microeconomics and macroeconomics. It evokes us to rethink: what is the purpose of economics? What is the purpose of money system? What is the value system? How can we incorporate physics in economics? What is the role of theology in economics? Can we tame the unruly human greed?... I submit to search answers in the Power of Living Water.

Matthew Yen

ISBN 978-1-68517-176-6 (paperback)
ISBN 978-1-68517-177-3 (digital)

Christian Faith Publishing
832 Park Avenue
Meadville, PA 16335
www.christianfaithpublishing.com

Printed in the United States of America

Footprint in the Sand

One night I dreamed a dream.
As I was walking along the beach with my Lord.
Across the dark sky flashed scenes from my life.
For each scene, I noticed two sets of footprints in the sand,
One belonging to me and one to my Lord.
After the last scene of my life flashed before me,
I looked back at the footprints in the sand.
I noticed that at many times along the path of my life,
especially at the very lowest and saddest times,
there was only one set of footprints.
This really troubled me, so I asked the Lord about it.
"Lord, you said once I decided to follow you,
You'd walk with me all the way.
But I noticed that during the saddest and
most troublesome times of my life,
there was only one set of footprints.
I don't understand why, when I needed You
the most, You would leave me."
He whispered, "My precious child, I love
you and will never leave you
Never, ever, during your trials and testings.
When you saw only one set of footprints,
It was then that I carried you."

This book is a footprint on how God led me through the beaches, mountains, and valleys, on this earthly journey. I want to dedicate it to God; my parents, Hong-Da and Mann-Hwa; siblings, Judy, Teresa, and Esther; my wife, Elaine, my children: Richard, Jonathan, Irene and their families, as well as people whose love inspired, encouraged, sustained and supported me.

I am extremely grateful to experience God's guidance as stated in Isaiah 43:19, "See, I am doing a new thing! Now it springs up; do you not perceive it? I am making a way in the wilderness and streams in the wasteland."

CONTENTS

LIST OF FIGURES *

---

* Note: Decimal figure number is used here *only* but not inside the book. For examples: 1.1 Cognition Classification is for Figure 1 in Chapter One. A.1 Water in a 3D Model is Figure 1 in Appendix A.

LIST OF TABLES**

** Note: Decimal table number is used here *only* but not inside the book. For examples: 1.1 System Similitude is for Table 1 in Chapter One. B.1 An Ongoing Debate is Table 1 in Appendix B.

NOTES TO READERS

Living Water assumes the perspective of "scientia ancilla theologiae". Such exposition was popular during the time of *Christendom* after Thomas Aquinas (1225–1274) and lasted through the European Renaissance of the 15th and 16th centuries. This writing style covers a broad range of relevant subjects that may hinder readers who could be benefitted from this book. Therefore, I am summarizing key points of motivations and background to guide readers at large:

Key Points:

1. The purpose of this book is to present a perspective of economics to include divine provisions beyond rational methodologies. We cannot overlook God's loving provisions, such as: water and air, in relation to human activities. Since Adam Smith's time, economics focused on the law of self-interest and, more recently, game theory and mathematical modeling. Nonetheless, we cannot exclude the acts of God from our daily lives, let alone economics. The book Living Water is an attestation of my life's purpose, goal, mission and ministry.

2. A gas model of the market empowers people to visualize Adam Smith's 'Invisible Hand' via the construct of Price-Volume (PV) and Temperature-Entropy (TS) conjugate variables of thermodynamics.

3. Spiritually, God breathed air into Adam's nostrils, giving him the breath of life in Genesis 2:7. Also, Jesus breathed air on disciples and gave the Holy Spirit in John 20:22. Therefore, the gas model is a practical manifestation

how humans behave and function in society via market economy.

4. The term of entropy was coined by Clausius in 1867 for the study of thermodynamics. Gibbs, Boltzmann developed the 'microscopic' understanding of entropy via statistical mechanics around 1917. While quantum mechanics was developed afterwards: Schrodinger developed the wave equation in 1926 and Heisenberg formulated the uncertainty principle in 1927. Both thermodynamics and quantum physics pointed out the stochastic or probabilistic characteristic of many natural phenomena.

5. Classical economists studied economics as a sociological phenomenon. Although Adam Smith advocated for the self-regulating principle based on the observation of steam engine flyball governor. Inspired by scientific advancements, neo-classical economists began to study economics with the attempt to develop mathematical models and rational constructs as a natural phenomenon. The stochastic nature of economics triggered a school of scholars to apply thermodynamics and quantum mechanics in the field of economics. Such endeavor provides us tools to analyze economics in objectively instead of individual perceptions or personal preferences. This book briefly summarizes the historical and scientific background of such an endeavor. Furthermore, Living Water elaborates beyond to discuss the philosophical and theological implications.

6. Regardless of modern theoretical development in economics, the influences of Adam Smith and Karl Marx are undeniable in recent history. The tension between China and Taiwan has been a reality since I was born seventy some years ago. People forgot about the log-jammed political situation was rooted in the grid-lock of the debate of capitalism and communism. People of China, Taiwan, USA, Russia, Ukraine… all suffer from this collective 'Amnesia'.

7. Economic theories must be backed with proper government structure and strategies. Rational methodology failed

to bring peace in history. It only makes sense for us to revisit Protestant Ethics advocated by Max Weber - Micah 6:8b—Do justice, love mercy and walk humbly with God. Christian ethics can only be upheld by honoring the tripartite government outlined in Isaiah 33:22: "For the Lord is our judge, the Lord is our lawgiver, the Lord is our king; it is he who will save us."

I invite readers of all background to join a journey searching for the unfathomable meaning of God's marvelous design, i.e.: air, water, light, spirit, and recent human history (His-story). This book addresses interests of diverse reader groups. Below is a table to guide readers to delve in the book with respective chapters:

Chapter / Reader interests	1	2	3	4	5	6	7	A	B	C	D
Economics	x	x	x	x	x			x	x		
Science & technology	x	x	x	x	x			x		x	x
General	x	x			x	x	x	x	x		
Spiritual	x					x	x				
Philosophy	x					x	x	x	x		
History	x	x				x	x	x	x	x	x
Politics & sociology	x	x						x		x	

PREFACE

Be warned, my son, of anything in addition to them.
Of making many books there is no end,
and much study wearies the body.

—Ecclesiastes 12:12 (NIV)

When I was in elementary school, the first thing I do in the morning was to help my mother to fire up a coal cylinder for cooking. Then I would go to the market to buy fresh vegetables and meat for Mom to cook for the day. When I was a teenager, my father decided to buy a refrigerator. It was a wonderful appliance that freed us from going to the market every morning. Like most boys, I was amazed by cars, ships, airplanes, and things that can move by themselves. Later I learned that all those moving things are powered by fuels or batteries. Fire, power, heating, and cooling are essential in our daily life.

In 1974, I went to the University of Idaho, knowing nothing about statistics, numerical methods, transport phenomena, computer drawing, etc. My master thesis is an equation of state (EOS) for air. In the course of transport phenomena, I learned about various forms of an equation of changes (EOC) for fluids, heat transfer, and mass transfer. Even thermodynamics was only limited to textbook knowledge. It was a divine provision that I worked with Drs. Stewart and Jacobsen on the correlating data for the equation of state (EOS) for air using multiple regression programs developed by them. It was a remarkable year of learning and thinking.

1

Dr. Steward recommended me to Purdue University as a research assistant at CINDAS, formerly known as TPRC, under the supervision of Dr. YS Touloukian. I helped with the data compiling for thermoelectric properties. Dr. Touloukian introduced me to the field of irreversible processes. During that time, I read a book *Introduction to Thermodynamics of Irreversible Processes* by Prigogine with interest and did my doctoral dissertation on a related topic.

My practical training ended up in the refrigeration industry. Afterward, I took interest in computer science and learned about neural network, artificial intelligence, and computer integrated manufacturing. In 1989, I was recruited to industrial technology at Fresno State University. The department is housed underneath the college of agriculture.

It was almost forty years later that I realized that Prigogine was a 1977 Nobel Prize winner in physical chemistry. He was an advocate of interdisciplinary research by applying physical science principles in social science. On my sabbatical year 2011, I read his book from *Being to Becoming*. It rekindled my interest in how to apply the equation of state and equation of change in social sciences. I realized that both EOS and EOC were developed in abstraction and isolated from society. Nonetheless, I tinkered with the idea of formulating EOS in economics.

Ten years ago, my son, Jonathan, asked me a question: "Daddy, is there any other way to determine the calories of foods besides burning them?" His question intrigued me to related food science in terms of thermodynamics. To my knowledge, the only way is to use calorimeter macroscopically, and there is no microscopic formula for such calculation according to their molecular constituents. Such inquiry prompted me to explore Willard Gibbs's contributions in physical chemistry in terms of internal energy, enthalpy, free energy and chemical potentials, etc. The work I have done on the equation of state for air at the University of Idaho came back to me. Not satisfied with those abstract concepts, I began to study Gibbs's life. I was impressed by his profound influences in the field of economics via Irving Fisher, Edwin Bidwell, and Paul Samuelson.

Not until February 2021 that I realized the demand and supply curve is a type of EOS and has the form of an ideal gas model. It daunted me that price elasticity in microeconomics has an underlying assumption of property of solid. Then I began to experiment with the idea to correlate compressibility factor, Z, to the price elasticity. I concluded that a gas model for the market economy is much more appropriate. I recognize Price-Volume diagram is not only an analogy of the law of supply and demand, which is widely used in microeconomic, but also offers an opportunity to compile compressibility products in all levels of the market (e.g., gas, real estate, airline, and computer). Furthermore, I postulate that "the invisible hand" coined by Adam Smith can be expressed via the TS diagram which offers a valuable tool for macroeconomic research and exploration.

It is an ambitious endeavor to bridge the field of economics and physics. Mimkes and his colleagues have researched the field of physical economics extensively. According to Mimkes, French medical doctor Francois Quesnay (1694–1774) pioneered this field by relating the production process as a circuit to the circulatory flow of human blood. Nonetheless, the term of *Econophysics* is too esoteric and alien for people who are interested in the subject of economics. On the other hand, people struggle to differentiate microeconomics and macroeconomics in the field of economics. To overcome the semantic barrier, I choose *living water* to provide a holistic perspective of microeconomics and macroeconomics: the analogy of circulatory flow of blood, the concept of liquidation and gasification, the essence of money flow and goods flow, the spiritual teaching of mastering economy instead of being enslaved by treasures, etc.

The first chapter addresses the need for a holistic perspective of microeconomics and macroeconomics, the issues of rational attitude in social science, and the methodology of metaphor and analogy. The second chapter relates the law of supply and demand to the equation of state based on the ideal gas model. Market elasticity is contrasted with the proposed market compressibility as a new concept in microeconomics. Chapter 3 relates the production cycle and business cycle in the framework of the Carnot cycle. The duality of PV diagram and

TS diagram offers an integrated framework for microeconomics as well as macroeconomics.

Chapter 4 explores the potential applications of PV and TS diagrams in macroeconomics. The effects and impacts of macroeconomic activities performed at constant-price, constant-volume, constant-temperature, and constant-entropy processes were explained.

Chapter 5 discusses the market growth and decline in terms of EOS and EOC. It discusses opportunities for applying quantum techniques, artificial intelligence, and neural networks. It also addresses challenges as the EOC is also governed by information and intelligence beyond physical world. The introduction of the Shannon entropy is a key to discuss the socioeconomic issues from philosophical and spiritual perspectives.

Chapter 6 uses a garden metaphor to discuss economics from a spiritual teaching perspective. This approach elevates the discussion of economics beyond the rational norm. Nonetheless, I believe that economics cannot be isolated from people and the acts of God.

Chapter 7 points out that history is a river of life living water. Both water and money are essential media in society. The characteristic of phase transition from the liquid phase to the steam phase is God's wonderful design. Steam has been called living water and powered human society physically, economically, and spiritually.

Appendix A summarizes the three laws of thermodynamics and three laws of econophysics by Mimkes. They correspond to the three laws stated by Adam Smith: the law of self-interest, law of competition, and law of supply and demand.

Appendix B points out that it is time to look beyond the dichotomy of capitalism and communism. It summarized a short list of people who influenced our generation. They are thinkers whose thoughts are still debated to this day, namely, Adam Smith and Karl Marx. People struggle with the ideology of large government or small government. Nonetheless, a schematic TS diagram offers an opportunity to guide us in a platform of continuum despite the senseless debate.

Appendix C is a summary of the irreversible processes. I took the liberty to extrapolate from physical chemistry to economics as a reference model to comprehend the phenomena of economics. Prigogine, a Nobel laureate, encourages applications of discovery in natural science in social science and advocates interdisciplinary research.

Appendix D is a summary of the art of idealization used in the fields of physical science and engineering. Idealization with proper assumptions helps us to simplify a complex system and come up with resolutions via imagination.

CHAPTER 1

Living Water Economics

He who believes in Me, as the Scripture said, 'From his innermost being will flow rivers of living water.'

—John 7:38

Hanauer (2019) was an early investor of Amazon company and a successful businessman. In a TED Talk on "The dirty secret of capitalism—and a new way forward," Hanauer stated:

> We need a new economics. So, economics has been described as the dismal science, and for good reason, because as much as it is taught today, it isn't a science at all, in spite of all of the dazzling mathematics. In fact, a growing number of academics and practitioners have concluded that neoliberal economic theory is dangerously wrong and that today's growing crises of rising inequality and growing political instability are the direct result of decades of bad economic theory. What we now know is that the economics that made me so rich isn't just wrong, it's backwards, because it turns out it isn't capital that creates economic growth, it's people; and it isn't self-interest that

promotes the public good, it's reciprocity; and it isn't competition that produces our prosperity, it's cooperation. What we can now see is that an economics that is neither just nor inclusive can never sustain the high levels of social cooperation necessary to enable a modern society to thrive.

He identified three misassumptions in neoliberal economics:

8. The market is an efficient equilibrium system.
9. The price equals its value.
10. A behavioral model of *Homo economicus* of being selfish, rational, and self-maximizing.

Hanauer further summarized five principles for new economics:

1. The market is a garden, not a jungle.
2. Inclusion creates growth.
3. Corporations exist for the community beyond the shareholders.
4. Greed is not good.
5. The laws of economics are a choice.

Hanauer's talk outlined some basic issues regarding the science of economics. He also pointed out we need a new economics that people create growth, not capital. Recently, scholars have approached economics from various areas (e.g., psychology, mathematics, and physics) all with a *rational* methodology. Hanauer's dissatisfaction with neoliberal economics is shared by people such as Warren Buffet, Charle Munger, and many practitioners in business and industries (see appendix B).

Science, Technology, and Rationality

The process of science and technology development is mostly based on a set of attitudes and orientations that are known as *rational*. However, in the face of capricious weather and a runaway economy, people find those rational tools or techniques limited. Rudi Volti (2017, p. 13) commented on the rationality:

> Rationality implies objectivity, coolness and detachment are part of the rational to understanding and changing the world. Guided by a rational outlook, scientific inquiry and technological application are usually based on abstraction or isolation of the part of the natural world that is being studied or manipulated. This is not always a good thing, for it can produce a sharp separation between the individual and the rest of the world. The scientist or technologist stands apart from the system that is being studied and manipulated, resulting in a kind of tunnel vision that may ignore the larger consequences of gaining and applying knowledge. For example, in discovering a genetic marker for a serious disease, a researcher might not consider potential abuses of that discovery, such as insurance companies refusing coverage of people with that marker.

Economics is a field that cannot afford strict abstraction and isolation. Therefore, it is proper to consider ways and means beyond the rational methodology of hard sciences. The contemporary rational methodology includes mathematical modeling, simulation, statistical analysis, neural networks, artificial intelligence, and big data.

Parable, Metaphor, and Analogy

It is the glory of God to conceal a matter and
the glory of kings to search it out.

—Proverbs 25:2

Economics is a field filled with the unknown as shown in the Johari window (figure 1).

Johari Window

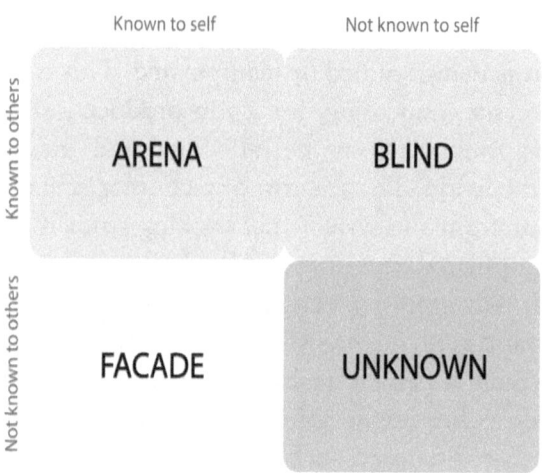

Figure 1 Cognition Classification—Johari Window as a guide for research. Economics is an area filled with unknown phenomena.

The rational methodology is not the only way to explain or understand the unknown area. When Jesus teaches, He often uses parables, stories to help people understand by *relating the unknown to the familiar things, objects, or characteristics in the open area.* Therefore, metaphor and analogy are viable methodologies to illuminate the unknown area such as economics.

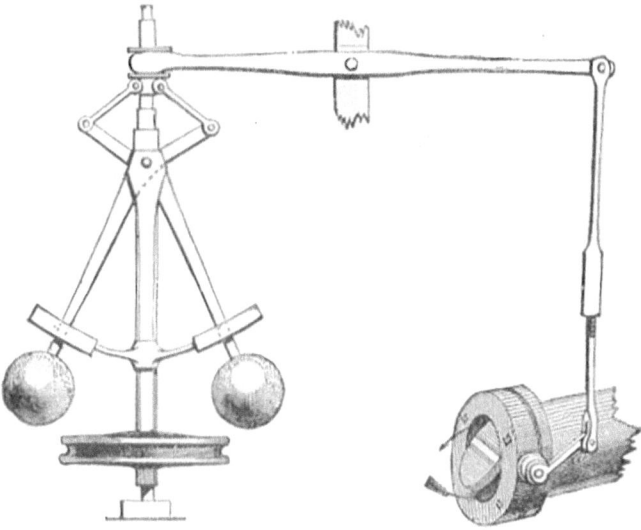

Figure 2 *Steam Engine Flyball Governor*—Change in the rotational speed of the vertical shaft causes the two balls to move up and down and control the linkage that opens and closes the throttle of the engine. Adam Smith adapted the feedback principle as the laissez-faire (French: allow to do) policy of minimum governmental interference in the economic affairs of individuals and society.

Volti (2017, pp. 11–12) construes that the Scottish economist Adam Smith (1776) developed his economic principles using the feedback principle of a steam engine flyball governor as a metaphor. The feedback principle of steam engine flyball governor (see figure 2) is a method of controlling a system by reinserting in the results of its past performance. Volti (2017) stated:

> During the late eighteenth century, the feed-back principle offered a suggestive metaphor for the workings of the economic system: instead of being guided by a centralized authority, an economy might best be organized through the operation of a self-regulating market, with the actions of independent buyers and sellers providing the

feedback. Thus, when buyers wanted a particular commodity, its price would be high, motivating sellers to produce more of it. If the price were low, less would be produced. In a similar fashion, an increase in production would cause the price of a commodity to fall, so more of it would be purchased, while a drop in production would cause the price to rise, leading to a reduction of purchases. In this way, the actions of buyers and sellers in the market provide a feedback mechanism through which supply and demand are supposedly brought into equilibrium.

Thus, *living water*—steam has not only powered steam engines but also powered economics in history, particularly, during the first Industrial Revolution in the sixteenth century. Financial terms such as *liquidation* and *currency* have been widely used in daily life. Liquidation has a connotation that money is *vaporized* or *gasified* when it is exchanged for assets, goods, or services. Currency has the connotation that money has the property of *fluid flow*. Such a concept is not surprising since both money and fluids are media in systems to facilitate the transport of energy, goods, services, etc. Figure 3 summarizes the properties of ideal money. Historically, money has evolved with more fluidic properties (see figure 4). Aside from the *fluid* property of currency, *idealization* is commonly adopted during theoretical development (see appendix D). John Nash postulated Ideal Money in early 1960. The ideal gas model and Carnot cycles were products of such *idealization*.

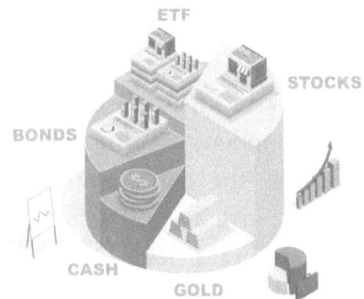

Properties of the Ideal Money	
Durable	Durable to circulate
Portable	Easy to transport
Divisible	Can be divided in small unit
Stable in value	Limited supply to keep value and avoid inflation

Figure 3 *Properties of Ideal Money*—Durability, portability, divisibility and stability are desirable properties of an ideal money system.

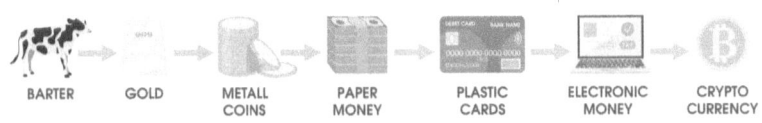

Figure 4 *History of Money*—Evolution of money indicates currency system is moving toward the ideal money system with more fluid characteristics in the future.

Econophyiscs—The Marriage of Economics and Physics

Richmond et. al (2013) have briefly summarized several renowned scholars who were interested in both physical sciences and economics, such as Aristotle (384–322 BC), Nicolaus Copernicus (1473–1543), Sir Isaac Newton (1642–1727), Blaise Pascal (1623–1662), and Edmond Halley (1656–1742). In the past century, scholars attempted to demystify economics from various perspective. Econophysics is an emerging school endeavor to integrate economics and physics, specifically thermodynamics. Samuelson (1970) explicitly acknowledged the maximum principles in analytical economics were due to the influence of the classical thermodynamic methods of Gibbs. Samuelson was the sole recipient of the Nobel Prize in Economics in 1970, the second year of the Prize. Georgescu-Roegen (1971) pioneered applying the thermodynamics framework to study economics.

The relationship between Samuelson and Gibbs is obscured from the general public mainly because of modern classification of disciplines. Albert Einstein praised Gibbs as "the greatest mind in American history." Gibbs had collaborated with Rudolf Clausius, James Clerk Maxwell, and Ludwig Boltzmann in the fields of statistical mechanics, thermodynamics, and mathematics. His doctoral student Edwin Bidwell Wilson was the mentor of Paul Samuelson, the 1970 Economic Nobel Prize awardee. Thus, Paul Samuelson sought to apply mathematics scientific principles in the field of economics.

Another doctoral student Irving Fisher was the first PhD in economic science at Yale University. Fisher was one of the American neoclassic economists. After the Great Depression of America, Fisher's neoclassic economy yielded to Keynesian macroeconomic thoughts advocates government fiscal and monetary policies to regulate the business cycles instead of self-regulating by markets. Samuelson's neoclassical economics focused on microeconomic market profit maximization while Keynesian macroeconomic thoughts seek to regulate aggregated demand and supply.

Gibbs and his contemporary Boltzmann endeavored the development of statistical thermodynamics. Their influences in the field of economics were at the infant stage. Rational methodology and mathematical analysis dominated in the academic fields following the progressive culture of the time. Nonetheless, we may relate Gibbs's thermodynamic quantities in terms of economic systems and thermodynamic in Table 1 aside from Samuelson's mathematic analysis.

Table 1 *System Similitude*—Analogous comparison of concepts in economic systems with relevant terms in mechanic and thermodynamic systems.

Domains/ characteristics	*Economic systems*	*Mechanical systems*	*Thermodynamic systems*
Intrinsic	Labor, quality, credit, creativity, motivation, assets, GDP, GNP, etc.	Potential energy	Latent heat, work, and chemical potential
Extrinsic	Currency circulation, information, communication, goods flows, etc.	Kinetic energy	Heat and mass fluxes via *transport phenomena* or *irreversible processes*

Mimkes (2006) applied the laws of calculus to formulate *econophysics*. The pros and cons of this approach are as follows:

Pros

1. Physics is rational, quantitative, and objective.
2. Metaphorical similarity between the thermodynamic cycles and economic cycles via concepts of energy, power, and work and familiar media, such as water and air.
3. Scalable because both macroscopic and microscopic frameworks are well established.

Cons

1. Abstract concepts and jargon, such as temperature and entropy.
2. Economics is an open system that involves people while physics and psychology do not mix.
3. The analogy between economics and thermodynamics has a narrow window for operational discussions.

Though econophysics has several limitations due to the rational attitude of isolation from the reality and abstraction construct, nonetheless, the ecological water cycle sustaining lives and economy is more like a fact than a metaphor. Therefore, there are certain noteworthy aspects, namely the ideal gas model and related construct.

Living Water—A Holistic Perspective of Microeconomics and Macroeconomics

Semantics and *rhetorical* are essential in social sciences. Terms used in econophysics, entropy, equations, etc. are alien to social science students. Living water economics may be more appealing to people instead of referring to it as econophysics. The term *living water* has multiple connotations. *Steam* is referred to as *living water* because the latent heat in the water powered the steam engine. Economically, the feedback principle of flyball governor for a steam engine inspired Adam Smith to publish his well-known *Wealth of Nation* as alluded to previously. Furthermore, water is a life-sustaining substance blessed by God. Spiritually, living water is referred to as blessings from God in daily life. Figure 5 illustrates how an ideal gas model is applied in the field of thermodynamics and economics.

PV and TS Diagrams

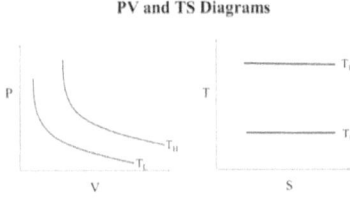

Physical science - Thermodynamics	Social Science – Economics
P- pressure	P- price
V- volume	V- volume
T - temperature	T – interest, trust, passion, confidence.
S – entropy, uncertainty, probabilities	S - uncertainty, options, possibilities, probability, production, capability, assets, resources, degree of freedom, etc.

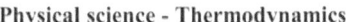

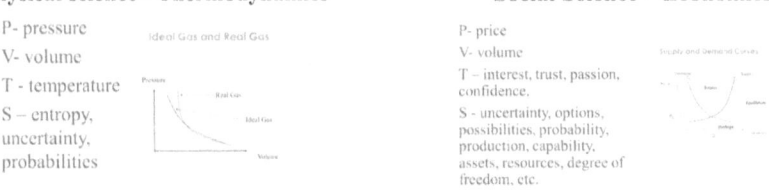

Figure 5 *Ideal Gas Model*—Ideal gas concept was hypothesized to study steam known as Living Water. The mathematic model is the base to study thermodynamics in physical science. Same model is also applied in economics to study the law of supply and demand.

Figure 6 shows that the PV diagram is a visual tool to characterize work performed by an engine system in thermodynamics while the TS diagram is a tool to show the amount of energy used to power the engine. In physical science, P stands for pressure, V is the volume, T is the temperature, and S is the entropy. According to Vocabulary. com, "The word *entropy* finds its roots in the Greek *entropia*, which means 'a turning toward' or 'transformation.'" The word was used to describe the measurement of disorder by the German physicist Rudolf Clausius and appeared in English in 1868.

"*A common example of entropy is that of ice melting in water.* The resulting change from formed to free, from ordered to disordered increases the entropy" (entropy, vocabulary.com). The word *entropy* (inside transformation) is best explained by the familiar phenomena of latent heat of ice transformed into water or water transformed into steam. The concept of latent heat was developed by Joseph Black in 1762. Nonetheless, the word *entropy* was introduced a hundred years later. Since the word *entropy* is a fairly new word in history and not commonly used in daily life, thus, people frown at the word and tend to mystify its meaning.

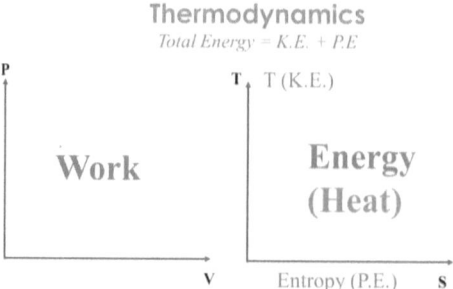

Figure 6 *Visual Aids for Thermodynamics*—PV and TS diagrams of the ideal gas model relate work and energy to study thermodynamics.

Georgescu-Roegen (1971), Mimkes (2006), and many econophysic scholars adapted the analogy of PV and TS diagrams for economics. Thus, figure 7 shows PV is a visual display of market performance or work accomplished while TS diagram is a visual tool to qualify the interest and trust of people as well as the resources engaged in the fundamental side of a market in terms of economic temperature, T, and economic entropy, S. In economics, P stands for price, V for volume or quantity, T is defined as economic temperature, and S is for the economic entropy, which will be furthered discussed in later chapters.

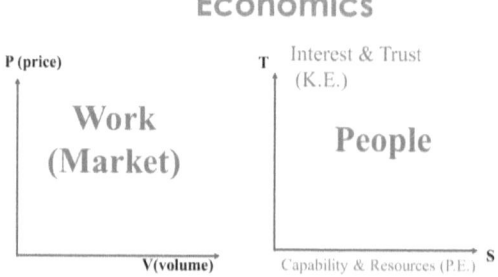

Figure 7 *Visual Aids for Economics*—PV and TS diagrams of the ideal gas model could be used in economics to relate work and people who power the market.

Microscopic and Macroscopic Perspectives

Generally, there are two levels of economics: microeconomics that primarily focuses on market supplies and demands while macroeconomics focuses on employment, economic growth, international trade, GDP, GNP, etc. For physical systems, there are also macroscopic descriptions and microscopic structures. See Table 2.

Table 2 *Level of Description*—Macroscopic and microscopic level descriptions are essential for both economics and physics.

Level of descriptions	Microscopic	Macroscopic
Economics	Business, markets, supplies, and demands	Regional growth, employment, GDP, international trade, aggregate supply and demand, inflation, monetary policy, etc.
Physics	Molecular, atomic structures, movements, interactions, etc.	System state variables relationships, changes, etc.

Furthermore, Table 3 is a summary of the difference between-macroeconomics and microeconomics. Table 3 summarizes the differences between macroscopic and microscopic approaches of physical properties. A benefit of the analogy between economics and thermodynamics is that it empowers us to explore economics at both macroscopic and microscopic levels.

Table 3 *Macroeconomics vs. Microeconomics*—Comparison of areas of study between microeconomics and macroeconomics.

Macroeconomics	Microeconomics
Analyze the economy as a whole	Analyze individual markets, industries and regional activities
Study aggregate economic behavior of groups, industries, e.g.: oil industry, insurance industry, etc.	Study economic behavior of individuals, companies, household on particular goods or market, e.g.: computer, car, housing, etc.
Deals with government policies, laws, regulations affect entire economy and society	Deals with the decision making of a company, household or individuals
Study aggregate variables, such as GDP, GNP, consumer index, unemployment rate, interest rate, exchange rate, aggregate demand and supply, etc.	Study the supply and demand in terms of prices, income, quantities, or volume.

Table 4 *Macrophysics vs. Microphysics*—Comparison between macroscopic and microscopic properties, scales and examples in physics.

	Macroscopic Properties	Microscopic Properties
Definition	Macroscopic properties of matter are the properties in bulk matter	Microscopic properties are the properties of the constituents of bulk matter
Visibility	Visible to naked eyes	Invisible to naked eyes
Unit of Measurement	In a scale that is visible, which includes cent-, kilo-,mega, etc.	In a scale that is invisible, which includes milli-,micro-, nano-, pico-, etc.
Examples	Volumes, pressure, temperature, density, etc.	Intermolecular forces, chemical bonding, etc.

Living water economics proposes an integrated economic platform for microeconomics and macroeconomics via the PV diagram and TS diagram (see figure 8).

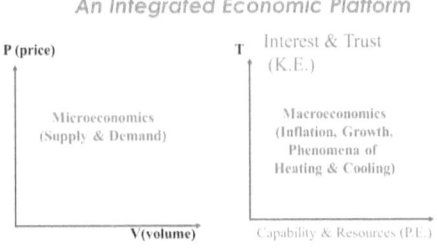

Figure 8 *An Integrated Economic Platform*—PV and TS diagrams provide an integrated platform for the analysis of microeconomics as well as macroeconomics.

The characteristics and topics for exploration in microeconomics and macroeconomics are summarized in table 5.

Table 5 *Equilibrium vs. Nonequilibrium*—Summary of the fields of study for microeconomics and macroeconomics in terms of equation of state in equilibrium systme and equation of change of nonequilibrium system.

	Characteristics	Microeconomics	Macroeconomics
Equation of State	Equilibrium, Time-Independent	PV domain analysis, Supply and Demand	TS domain analysis, Asset & Resources Management , Labor & Costs, Heating and Cooling
Equation of Change	Non-equilibrium, Time-Dependent	Tax, trades, competitions, weather, wars, pandemics, insurance, etc.	Inflation, Growth, Depression, Money Policy, Employment, Risk Management (Acts of God), GDP, GNP, etc.

Cartoon, Picture, and Movie

It is amazing that a cartoonist can capture the characteristics of an object and express them in a few pen strokes. Picture or still image with more details will help us to understand the structure of the object. While a video clip or movie will further our understanding of the behaviors of the object. This relationship is illustrated in figure 9. Economics is a subject that consists of characteristics, properties, and behaviors just like any object. In abstract terms, it can be summarized in laws, equation of state, and equation of changes. Such relationships are summarized in table 4.

Figure 9 *Object Example*—Eagle is an object can be studies via cartoon, picture, and movie. In this case, movie of eagle life, snapshot of eagle flying and line drawing of eagle flying. Likewise, economics can be studies via data analysis, equation of states and equation of changes.

Table 6 *Object Descriptions*—Comparison of tools used to study an eagle and economics via object concept.

Cognition levels\objects	Eagle	Economics
Characteristics	Cartoon	Data analysis
Properties	Picture	Equation of state
Behaviors	Movie	Equation of change

In the past century, the rapid development of computer and information technologies has propelled the field of knowledge representation. A major achievement is object programming and related applications. In object programming, all objects have a shared construct, namely, attributes, methods, or events. The introduction of the concepts of the equation of states and equation of changes may pave the way to study economics in terms of object constructs to harness the dynamic nature of economic activities.

The Spirit of Living Water

The ethos of the Living Water economy is best illustrated by comparing the Sea of Galilee and the Dead Sea in the Middle East (see Figure 10). Both bodies of water are fed by the Jordan River, yet the Sea of Galilee is beautiful, lush, full of fish, and teeming with life, while the Dead Sea is a lifeless desert. This economy existed without any economic theories, such as The Wealth of Nations by Adam Smith or communism by Karl Marx. When Adam Smith published The Wealth of Nations (1776–1783), America was in the midst of the revolutionary war with Britain (1774–1783). The thirteen colonies issued individual treasury notes without a unified currency system. Essentially, the Living Water economy of those days was based on trusted relationships, rather than eloquent economic theories.

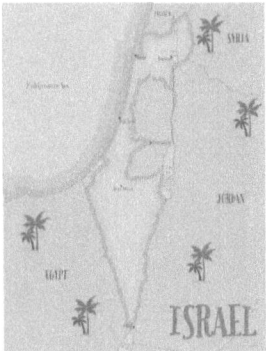

Figure 10 The Spirit of Living Water—Jesus used the disparity of the burgeoning Sea of Galilee and the lifeless Dead Sea to illuminate the Spirit of Living Water.

CHAPTER 2

Microeconomics and Equation of State

People often use emojis to express their emotions. Emoji is a snapshot of the state of a person. In economics, the market for trading is dynamic. Knowledge of goods value and availability is essential for both consumers and suppliers for price negotiation. Timely snapshots of the status of supply, demand, and price are valuable instruments. Mathematically, the snapshot is known as the *equation of state* (EOS).

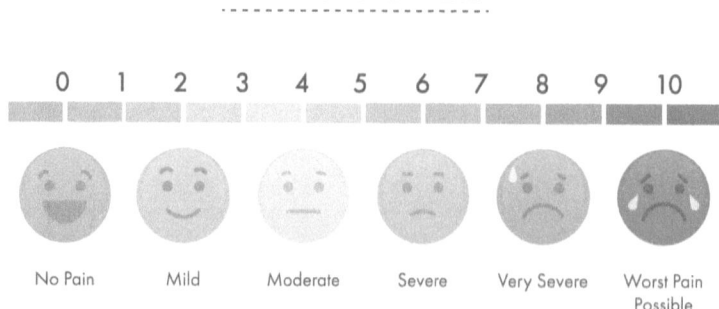

Figure 1 *Pain Scale*—Face expressions or emoji's reflect the state of a person's pain or emotion. Likewise, equation of state (EOS) is a snapshot of the market in economics.

Law of Supply and Demand

In 1776, Adam Smith published *the Wealth of Nations*. He stated three laws of economics: (1) law of self-interest, (2) law of supply and demand, and (3) law of competition. The law of supply and demand is widely used in explaining the behavior of economics. Figure 2 is an illustration of the law of supply and demand.

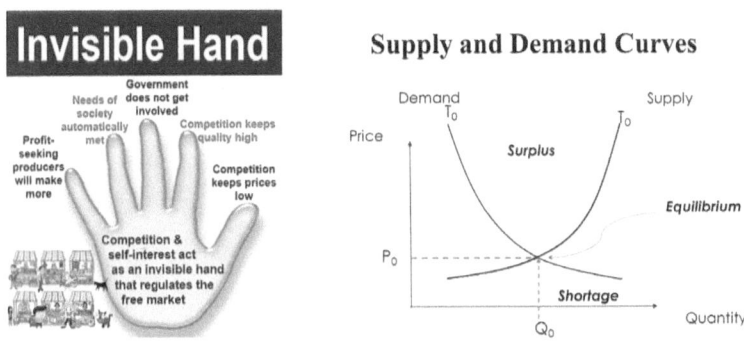

Figure 2 *Adam Smith's "Invisible Hand"*—According to Adam Smith the law of supply and demand indicates price is generally determined at the equilibrium point where the supply curve and the demand curve intersect each other. Adam Smith coined the term of 'invisible hand' for the law of supply and demand.

Scholars studied economics extensively using the law of supply and demand as the basis of microeconomics. Generally, the demand curve is mathematically a hyperbola. For simplicity, it is approximated as a line. Richmond (2013) modeled demand and supply functions in the form of a Boltzmann distribution (pp. 183–184). Nonetheless, economists have classified the demand linearly according to the slopes of demand curves as elastic, inelastic, and unit elastic according to the slope of price changes (see figure 3). The term *elasticity* implies that the market exhibits the behavior of an elastic solid as used in solid mechanics. This underlying assumption has not been challenged in the past. Nonetheless, with the advance of science, many scholars propose to model economic behaviors in terms of fluid and thermodynamics (see references).

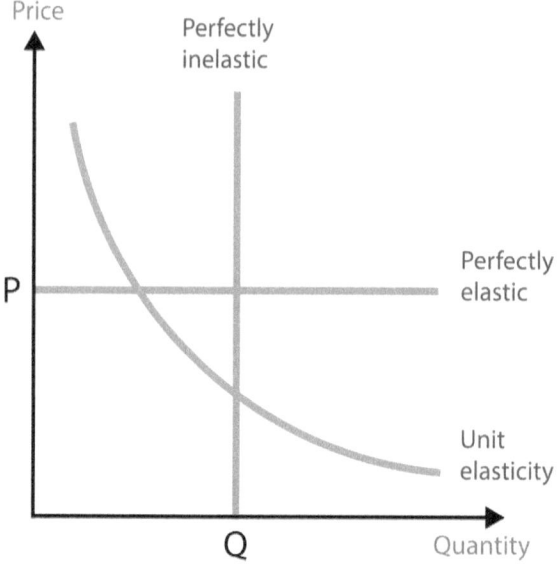

Figure 3 *Price Elasticity*—In economics market assumed to behaves in the sense as a solid with various elasticities. Price elasticity is considered as a property of the product or service.

Table 1 summarizes the analogies of state variables for economic and thermodynamic systems (Saslow).

Table 1 *Variable Similitude*—Summary of Analogies Between Economic And Thermodynamic Systems According to Saslow (1999)

	Thermodynamics	Economics
P	Pressure	Price
V	Volume	Volume, quantity
T	Temperature	Temperature (interest, credit, value, confidence)

Microeconomics—PV Diagram

To model economic behavior, it is best to begin with the ideal gas model. Benoît Paul Émile Clapeyron first stated the ideal gas law in 1834. Figure 4 shows that pressure and volume are related in hyperbolic form along the isothermal-constant temperature lines. Thus, if pressure and volume are known, temperature can be readily determined according to the empirical law: $PV = nRT$. Note that T is the absolute temperature. This is the base form for the equation of state for real gases, such as oxygen, nitrogen, and air.

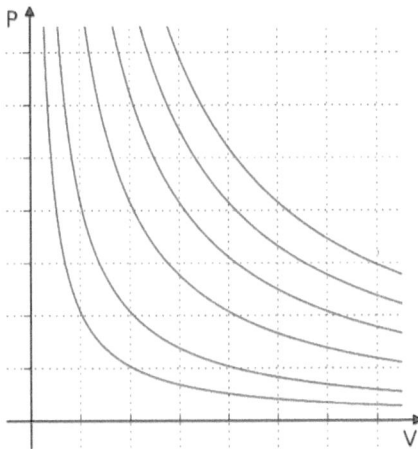

Figure 4 *Ideal Gas Law*—Pressure and volume of Ideal Gas related by the Equation of State PV=nRT

Real gases, such as oxygen, nitrogen, and air, will deviate from ideal gas relationship (see figure 5). To account for deviations, numerous terms have been included via experimental data and regression analysis as shown in Figure 6.

Ideal Gas and Real Gas

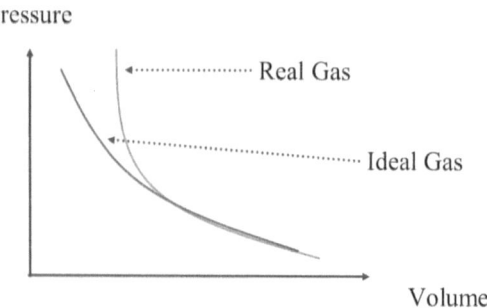

Figure 5 *Real Gas vs. Ideal Gas*—Real gas PV curve deviate from ideal gas law. (Economically, different product market will exhibit its own characteristic aggregated demand curve or supply curve from ideal demand and supply curves. Ideally a hyperbola)

Computational Models

Equation of State for Air

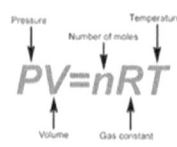

$PV=nRT$

Opportunities:

Computation Models for
Market Demand and Supply
Of Markets at All Scales,
e.g.: iPhone, steel, oil , etc.

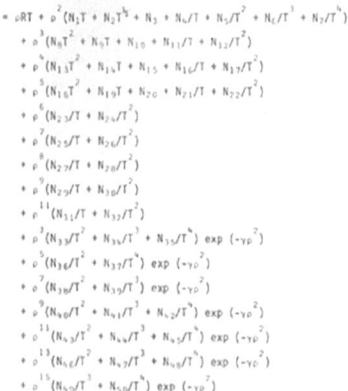

Figure 6 *Computation Models*—Equation of state for air has a large number of coefficients and parameters, which are determined via experimental data and multiple regression analysis. Likewise, it is feasible to employ big data and artificial intelligence (AI) to determine supply and demand curves for products (i.e., iPhone, steel, cars, and commodities).

Not only the demand curve exhibits the hyperbolic form similar to that of an ideal gas, but the supply curve also shares the same form in the opposite direction as shown in figure 7.

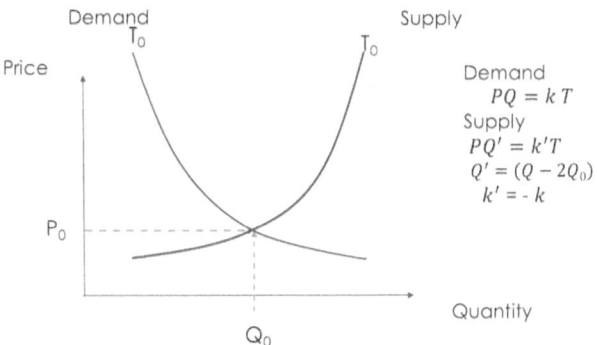

Figure 7 *Mathematic Models for Supply and Demand*—Supply curve share the same hyperbolic form as demand curve in the opposite direction.

It is a common practice in physical science to show the characteristics of various gases by the compressibility factor Z=PV/RT as shown in figure 8.

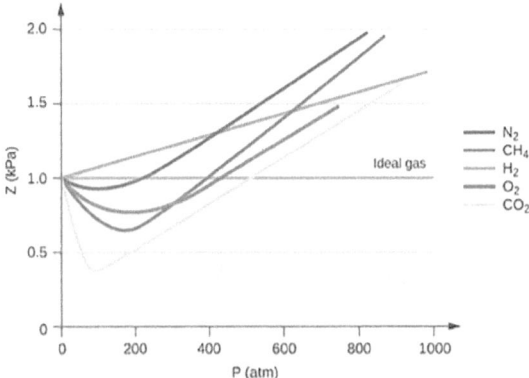

Figure 8 *Compressibility Factor*—Real gases are characterized with various compressibility factor, Z, along the pressure scale.

By assuming the market behaves like a gas model, it would be inspiring and revealing to construct a library of negotiation efforts (compressibility factors) for various products like figure 8 for different gases. Table 2 compares price elasticity, e, and market compressibility, Z. When the demand curve follows a perfect hyperbola, the price elasticity e = 1 and Z = 1. While in a market where price does not change with the demand such as the case of plentiful supply of water, e = 0. The market is perfectly elastic. When 1 > e > 0, the market is elastic and negotiable. Therefore, market compressibility factor Z < 1. When a market has limited supply such as the case in an auction. The price jumps up quickly. Price elasticity e rises higher. The market is stiff and less negotiable. Therefore, infinite > e > 1 and the market compressibility factor Z > 1 as in the case of less compressible gases.

Assuming an ideal market follows the hyperbola of PV = 10. Figures 9 is an example showing how the market compressibility factor, Z2 varying with volume V in a market with a linear demand curve. Horizontal line Z1 is the compressibility factor of an ideal market of the hyperbola. Figures 10 is another example showing how the market compressibility factor, Z2 varying with volume V in a market with a linear demand curve of different slope. Again, the horizontal line Z1 is the compressibility factor of an ideal market of the hyperbola.

It would be a great contribution that institutions may construct libraries of compressibility factors for different industries, such as agricultural, electronic, and health products and cars.

Table 2 *Price Elasticity and Market Compressibility*—Comparison of price elasticity, e, and compressibility factor, Z.

Price elasticity, e	e = 0 perfect elasticity (Invaluable); 1 > e > 0 Elastic market (abundant supply)	Unit elastic-ity: e =1	Infinite: perfect inelastic Valuable and nonnegotiable; Infinite > e > 1 Stiff market (short-age of supply)
Compressibility factor, Z	Z< 1 compressible, Negotiable market (abundant supply)	Z = 1 Idealized market behavior	Z > 1 Less compressible, negotiable market (shortage of supply)

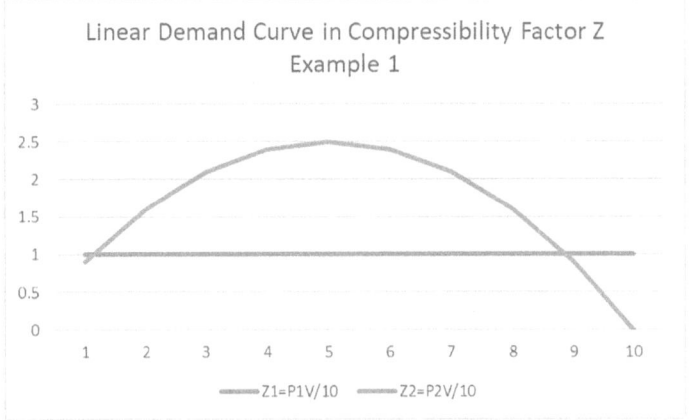

Figure 9 *Market Compressibility Example 1*—Compressibility Factor for Hypothetical Product A with a linear demand curve. In this case Z > 1 increases to a point then begin to decrease and drop to Z < 1.

The compressibility factor (Z) could be used as an index for the negotiation effort to reach an agreeable price between buyers and sellers. Therefore, Z could be a valuable index for business dealings.

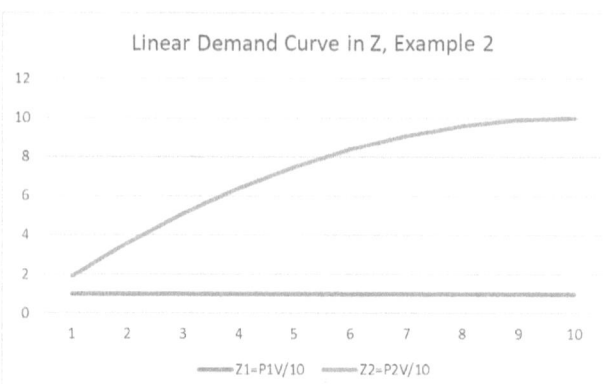

Figure 10 *Market Compressibility Example 2*—Compressibility Factor for Hypothetical Product B with a linear demand curve of different slope. Here Z > 1 increases as the volume increases.

Figure 11 is the compressibility factor of a perfect elastic market. Note that the market is linearly incompressible along the price line for a perfect elastic market. From a business standpoint, it implies that negotiation effort increases linearly. Figure 12 shows that a perfect inelastic market is also incompressible or not negotiable.

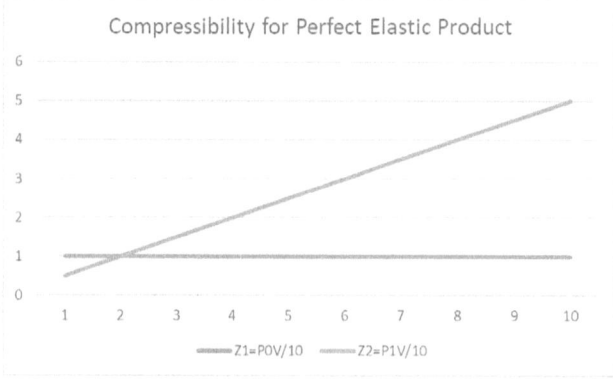

Figure 11 *Perfect Elastic Market Compressibility*—Compressibility Factor for a perfect elastic market. Negotiation effort increases even in a perfect elastic market.

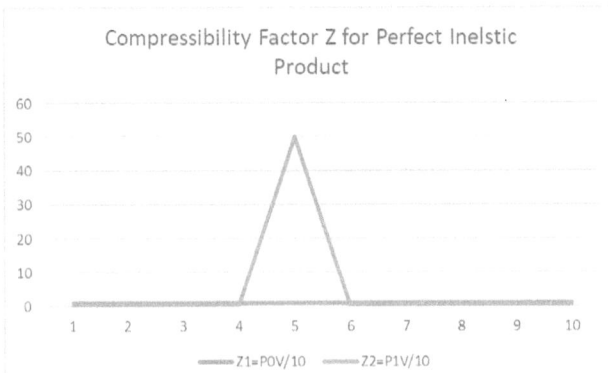

Figure 12 *Perfect Inelastic Market Compressibility*—Compressibility Factor for a perfect inelastic Product. No room for price negotiation in a stiff market.

In the stock market, PV diagrams provide an additional analytical tool. Figures 13–15 are snapshots of 2020 for three selected companies (i.e., ASML, Amazon, and Tesla). Linear elasticity is obtained via regression analysis based on the 2020 data.

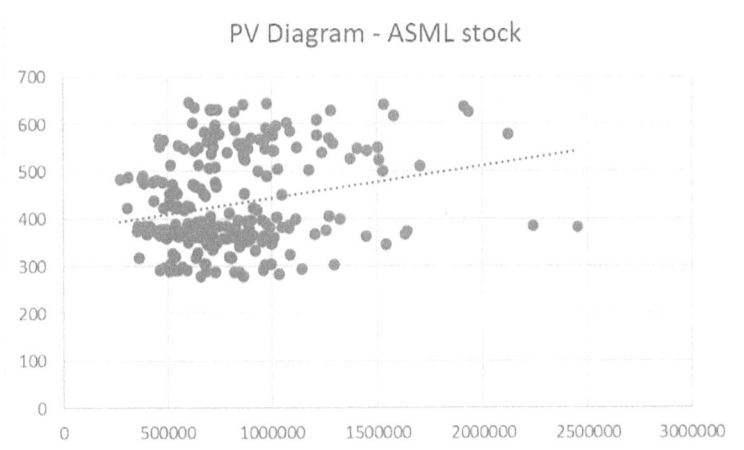

Figure 13 *PV snapshot of 2020 ASML Stock*—Note PV diagram is Not in the time domain. Therefore, the regression line represents the loci of the equilibrium points over the period of 2020.

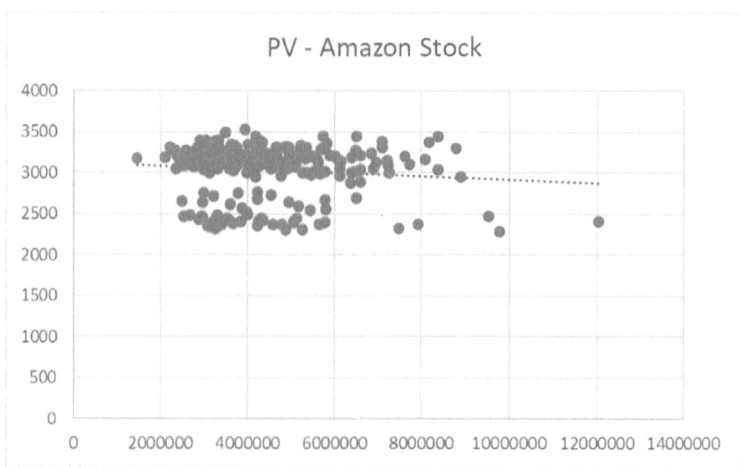

Figure 14 *PV snapshot of 2020 Amazon Stock*—Note that the regression line is the loci of the equilibrium points over the period of 2020, which resembles a line of a perfect elasticity.

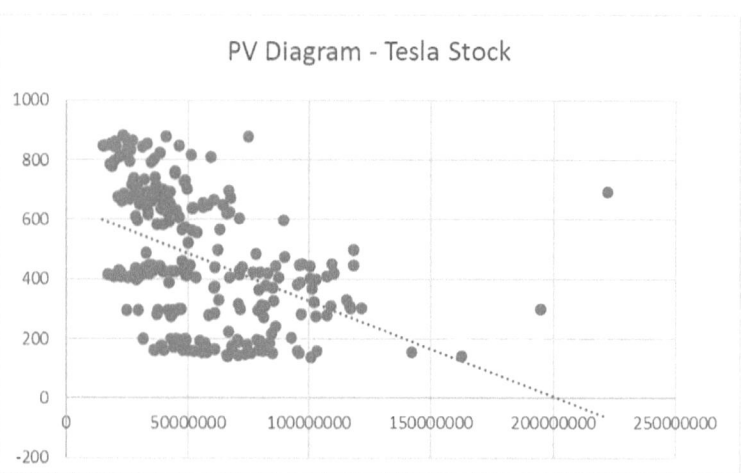

Figure 15 *PV snapshot of 2020 Tesla Stock*—The elasticity of the loci of equilibrium points indicates the market is sensitive to stock price fluctuation.

CHAPTER 3

Production Cycle and Business Cycle

And God said, "Let there be a vault between the waters to separate water from water." So God made the vault and separated the water under the vault from the water above it. And it was so. God called the vault "sky." And there was evening, and there was morning—the second day.

—Genesis 1:6–8

Production Cycle

In the agriculture society, growing crops and raising livestock on a farm were means to sustain lives. When the population increases, farm operations became complicated. Figure 1 is a schematic diagram to illustrate the farm-to-fork cycle.

Farm to Fork Continuum

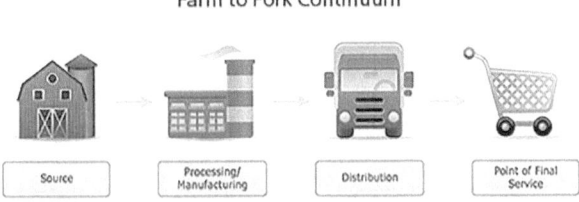

| Source | Processing/ Manufacturing | Distribution | Point of Final Service |

Figure 1 *Farm to Fork*—This is the most basic economic cycle to sustain people's lives.

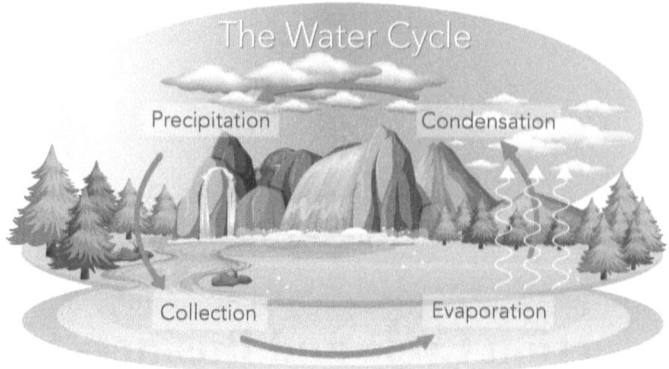

Figure 2 *Water Cycle*—Evaporation, condensation, precipitation and collection form a water cycle which is essential to grow crops and power the economy.

Whether to grow crops or to raise animals, sunlight and water are essential for farm operation and production. Therefore, the water cycle in figure 2 is a major component of the farm-to-fork agricultural operation cycle.

Now let us look at the steam engine cycle. Steam is referred to as living water to power industrial equipment and the economy. The study of water/steam properties led to the development of thermodynamics. It is also instrumental in the development of the ideal gas model. Figure 3 is a train powered by a steam engine.

Figure 3 *Steam Engine*—Steam is referred as Living Water to power train, electrical power plant, and numerous industrial equipment.

A steam engine is operated by burning coals to heat water to generate steam. High-pressure steam is used to push a mechanism to do the work. In thermodynamics, the product of PV represents work, w, performed by a mechanical system. The product of TS represents heat, q, transferred across the system to generate steam. According to the first law of thermodynamics, work and heat are interchangeable via a certain medium, in this case, the steam (c.f. figure 4 and appendix A).

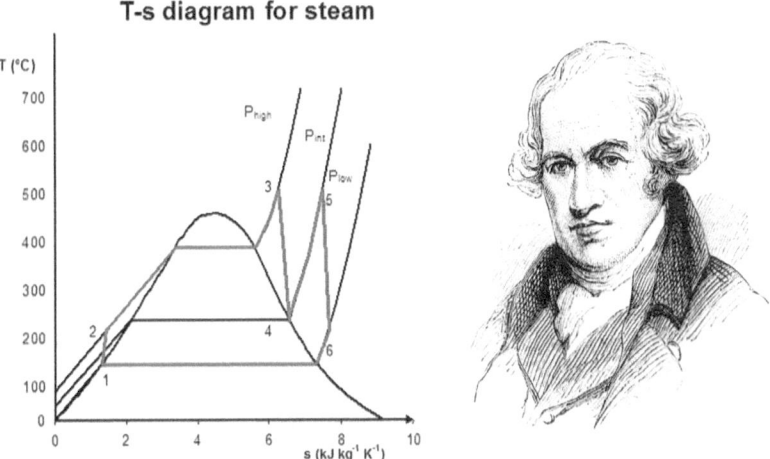

Figure 4 *TS Diagram for Steam*—James Watt's observation led to the study of the thermodynamics properties of water and steam. TS diagram diagrams are essential tools for designing engine, power plant and refrigeration systems.

The body of knowledge regarding heat, steam, thermodynamic properties, and related subjects not only triggered the first industrial revolution but also influenced thoughts in other fields of sciences (e.g., economics, communication, information science, politics, and social sciences in general). Table 1 is a summary of such historical accounts.

Table 1 *Key Milestones*—A summary of related historical events and key figures in an effort to describe complex systems in the field of thermodynamics, communication, information, and economics.

Year	Events	Key figures
1761	Latent heat	Joseph Black
1763–1775	Steam engine	James Watt
1759–1776	Concept of the *invisible hand* (Divine Providence—Hand of God), *laissez-faire* (let do)	Adam Smith
1824	Carnot cycles	Nicolas Léonard Sadi Carnot
1834	Ideal gas law stated	Benoît Paul Émile Clapeyron
1850	Introduced the word *thermodynamics* (θέρμη-δύναμις, meaning *heat power*)	James Joule
1862–1865	Introduced the word Entropy (εντροπία) and formulation Formulated the equation of state with state variables P, V, T, and S (entropy)	Rudolf Clausius
1870	Formulated statistical entropy from the microscopic approach as the sum of probability of all possibilities.	Ludwig Boltzmann
1948	Applied the concept of entropy in the field of communication as *uncertainty*	Claude Shannon
1971	Apply thermodynamic framework in economics	Georgescu-Roegen

The thermodynamic properties of water and steam have been researched by many people in the seventeenth and eighteenth centuries. Nonetheless, Benoît Paul Émile Clapeyron postulated the ideal gas law in 1834 to aid the study of steam and other gases. Figure 4 shows the PV and TS diagrams for an ideal gas.

The PV diagram is used to calculate the amount of work, w, of a system while the TS diagram is used to calculate the amount of heat, q, injected into a system. In physics and engineering, the PV and TS diagrams are conjugated tools to design a power engine or a cooler.

Math Mathematically, change of internal energy,

$$\Delta u = \text{heat}, q - \text{work}, w \quad (1)$$

Here, Δu represents the change of internal energy of a system. The state diagrams of the conjugate of P-V and T-S are often used to illustrate the performance of a mechanical system.

In thermodynamics, entropy is defined as follows:

$$\text{Entropy} = \text{latent heat/temperature} \quad (2)$$
$$S = q/T$$

Therefore, entropy represents the latent heat of the steam. Microscopically, heat absorbed by water increases the number of microstates. Boltzmann relates entropy to the probability of microstate.

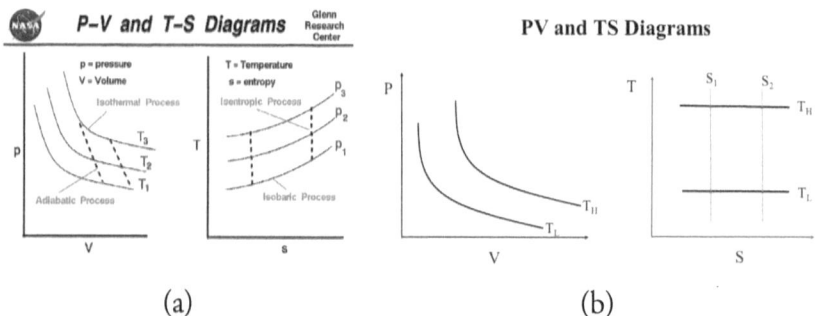

(a)   (b)

Figure 5 (a) *PV and TS Diagrams*—Rendered by NASA Glen Research Center. These are basic design tools for rocket engine.

Figure 5 (b) *Isothermal Lines and Equipotential Lines*—The isothermal (constant temperature) lines in PV and TS diagram simplify the study of thermodynamics. The PV diagram is a tool to examine the work performed by a heat engine. While TS diagram is a tool to depict the relationship between Work and Heat or Power.

The PV diagram is a tool to examine the work performed by a heat engine. While TS diagram is a tool to depict the relationship between Work and Heat or Power.

Note: The isothermal lines and isentropic lines in TS diagram of figure 4 (b) resemble the equipotential lines as in geographical elevation or gravity field. Likewise, the isentropic line (constant entropy) lines represent a potential filed. Such construct simplifies discussions of thermodynamic in familiar terms, i.e.: elevation and potential. That is the beauty and elegance of TS and PV diagrams.

Carnot Cycles

Nicolas Léonard Sadi Carnot in 1824 explored the upper limit of conversion efficiency of a theoretical heat engine or a refrigerator and proposed idealized Carnot cycles for heating and cooling. See figure 5 and figure 6.

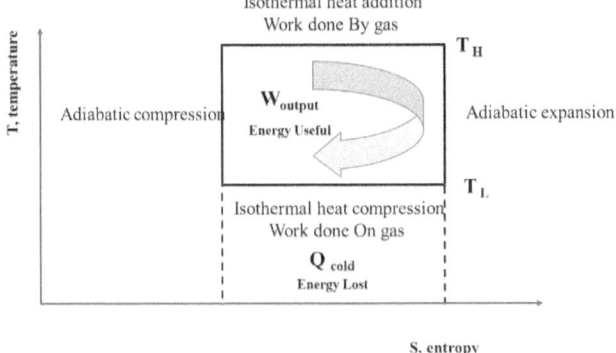

Figure 6 *(a) Carnot Power Cycle*—Note the direction of the cycle is 'clockwise'.

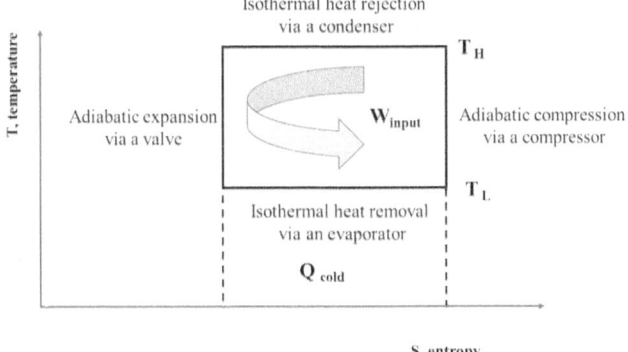

Figure 6 (b) *Carnot Cooling Cycle*—Note the direction of the cycle here is 'counter-clockwise'.

The Carnot cycle is a theoretical reference to guide the design of all types of engines and refrigeration systems. They are *idealized* via unrealistic assumptions to simplify complex systems (see appendix D, The art of idealization). Please note that Carnot cycles describe closed systems such as a steam engine and assume the system is at equilibrium while a water cycle as shown in figure 2 is an open system. An open system is constantly disturbed by external factors, such as the sun, wind, people, and animals. The water cycle is part of the larger field of meteorology. We all know how uncertain weather forecasting is.

In a lab, an open system may be studied with models and a set of the equation of changes (EOC) for pressure, volume, temperature, and entropy. But in real world, such EOC can only be simulated. To use the thermodynamic framework to model economic activities, we must clarify the following: *variables, equation of state* (EOS), and *equation of changes* (EOC).

Variables

Table 1 in chapter 2 established the analogies of pressure, temperature. The table is repeated here with the inclusion of entropy.

Table 2 *Variable Similitude*—Summary of analogies of state variables between economic and thermodynamic systems.

	Thermodynamics	Economics
P	Pressure	Price
V	Volume	Volume, quantity
T	Temperature	Temperature (interest, credit, value, confidence)
S	Entropy	Entropy (capability, degree of freedom)

The state variables of price and volume in a market are well understood. Temperature (T) and entropy (S) in economics demand explanations.

Economic temperature

In thermodynamics, temperature (T) is defined as the tendency of an object or system to spontaneously give up energy. The determination of thermodynamic requires extensive calibration in thermometry with instruments.

In economics, Mimkes defined *economic temperature* as an integrating factor. Bryant (2011) used the concept of kinetic theory assigned trading value as the *economic temperature*. Since economics is a field of social science, to define the objective temperature scale as the absolute temperature scale in physics is a vain effort. *Generally speaking, "temperature" in social sciences is much more sensible and feelable since we all have needs, wants, desires, likes, dislikes, and motivations.* The intensity or degree of interest, trust, belief, and faith is commonly reflected by attitude, words, expressions, or actions. For groups, the *temperature* toward certain products is sensible via data collection instruments (e.g., surveys, interviews, voting, and Internet platform).

Temperatures are essential variables in daily life, ranging from body temperature, weather temperature, cooking temperature, to lab temperature. Likewise, economic temperatures are essential for government planning, product and service marketing, employment, etc. The government may use GNP, GDP, and GNI to gauge the economic temperature. Corporations may use the Dow Jones index, Nasdaq index, Standard & Poor's (S&P) index, and market survey tools to estimate it. Individuals and households most likely determine the economic temperature by value. Regardless, the economic temperature is an index of interest and a force that drives economic activities.

From the psychological perspective, Maslow (1943) proposed the hierarchy of needs in his paper *A Theory of Human Motivation*. It is a visual illustration of the law of self-interest (see figure 7).

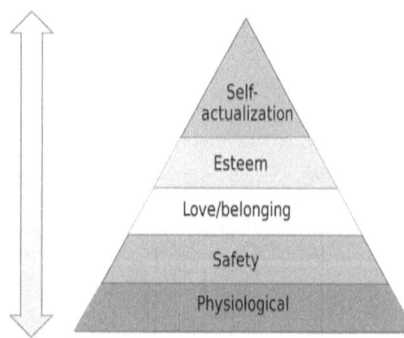

Economics driven by stratified human needs vertically integrated via banking & finance, information technology, and service sectors.

Figure 7 *Maslow's Hierarchy of Needs*—This hierarchy is a visual illustration of the law of self-interest, which is a simple model for market stratifications. Economics driven by stratified human needs vertically integrated via banking & finance, information technology,and service sectors.

This intensity and degree of need for a product or service can be a measure of the motivation, interest, or the (economic) temperature. In a market, $T = 0$ simply means there is no interest or motivation. On the other hand, a high frequency of transactions means there exists temperature in the market, and the number of transactions is the measure of the temperature. Maslow's hierarchy is not only used for individuals but can also be adopted for groups or organizations.

Economic entropy

Entropy (S) in thermodynamics is a state variable alien and abstract to many people. However, entropy change can be demonstrated by a commonly observed kitchen phenomenon:

Εντροπψ

熵

Shāng

Note: Entropy in Chinese is 熵, a word combined with fire (火) and business (商). 熵 has the connotation of incubation or brewing a business by gathering resources and talents to launch a business. Thus, entropy represents the potential or sum of all resources.

Figure 8 *Entropy Means Brewing a Business*—Heat absorbed by water increases its temperature and water transition into vapor phase as steam (living water). Entropy of water increases at the boiling point.

The amount of heat absorbed by the water droplet at the heating temperature is known as entropy change. Metaphorically, economic entropy is the number of resources to enable the *phase transition* of an economic entity or agent in the sense of an increased degree of freedom or capacity. For an orchard, the entropy can be the number of trees or acreage. For a manufacturing company, entropy can be an indication of its manufacturing capability (see figure 8). In all cases, it can be translated into the *dollar amount/degree*.

Figures 9 and 10 summarize the variables for thermodynamics in physical science and economics in social science, respectively. In physical science, variables are measured with instruments as a real number. In economics, variables are surveyed or collected. To make the analogy meaningful, economic temperature and economic entropy will adopt a scale of measurement in terms of interval or ratio (see figure 11).

Physical science - Thermodynamics

P- pressure

V- volume

T - temperature

S – entropy,
uncertainty,
probabilities

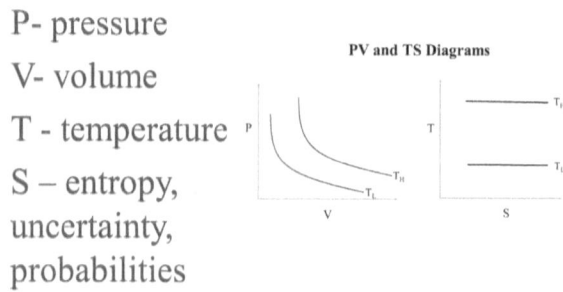

PV and TS Diagrams

Figure 9 *Thermodynamic Variables*—variables used in thermodynamics are measurable.

Social Science – Economics

P- price

V- volume

T – interest, trust, passion,
confidence,

S - uncertainty, options,
possibilities, probability,
production, capability,
assets, resources, degree of
freedom, etc.

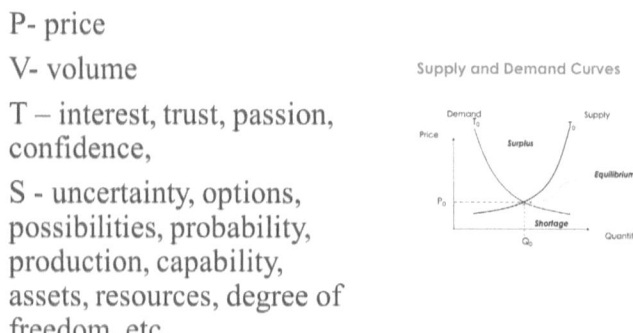

Supply and Demand Curves

Figure 10 *Economic Variables*—Variables in economics are surveyed.

LEVELS OF MEASUREMENT				
	Nominal	*Ordinal*	*Interval*	*Ratio*
Attributes	Named Variables			
Types	Qualatative		Quantitative	
Examples	'Hair color'	'Level of pain'	'Temperature'	'Height'
Mutually Exclusive Categories	X	X	X	X
Natural or fixed order	no ordering	X (ranking, ordering, or scaling)	X	X
Equal interval or spacing			X	X
Absolute zero so that ratio can be calculated				X

Figure 11 *Levels of Measurement*—Variables are surveyed as data according to scale measurement.

Environment temperature is critical for human living and habitation. Body temperature is an indication of a person's health. So is the economic temperature essential for society sustainability and wealth growth. Though there is no thermometry for economic temperature as in physical science, nonetheless, ranking and rating systems have been widely used by statisticians and data analysts in this Internet era. Google Search is also developed by eigenvectors. Those are viable tools to determine economic temperatures as well as economic entropy according to applications.

Macroeconomics and the TS Diagram

As the head and tail of a coin, PV and TS are two sides of an entity—PV represents the head, or the market side, and TS represents the tail, or the fundamental side. Therefore, the TS diagram is an ideal tool for the exploration of macroeconomics. By using the TS diagram, we can incorporate the psychological factors of individuals and groups into the discussions of economics as the economic temperature (T). The state of people's psychology is the driving force of the economy. It likens the kinetic energy (KE) of a ball. On the other hand, entropy is a measure of the production capability of individuals or groups. It likens the elevation of a waterfall or potential energy (PE) of individuals or groups. Therefore, the TS diagram is a chart for the health chart or fundamental side of a market, which has been lacking in the study of macroeconomics.

Carnot Cycles in Economics

With the understanding of the variable definitions, figure 12 illustrates the typical Carnot cycle in the TS diagram in economic. Note: the Carnot cycles may be a *heating* or a *cooling cycle*, depending on the direction of the media (i.e., money flow).

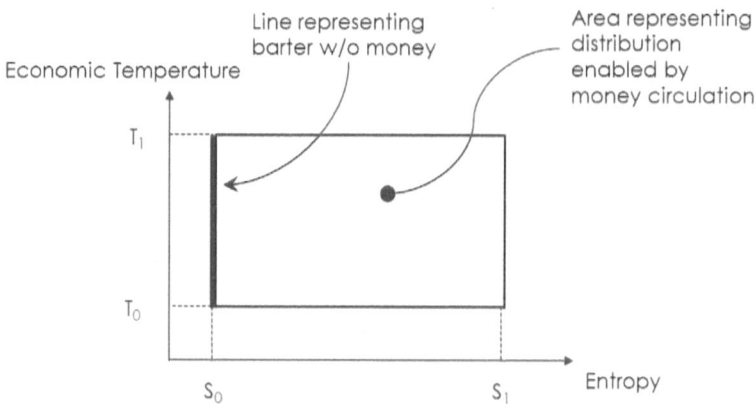

Figure 12 *Carnot Cycle Metaphor*—The Carnot cycle in a TS diagram may illustrate economic cycles.

Work Cycle

An idealized work cycle can be constructed to describe activities in economics in figure 13:

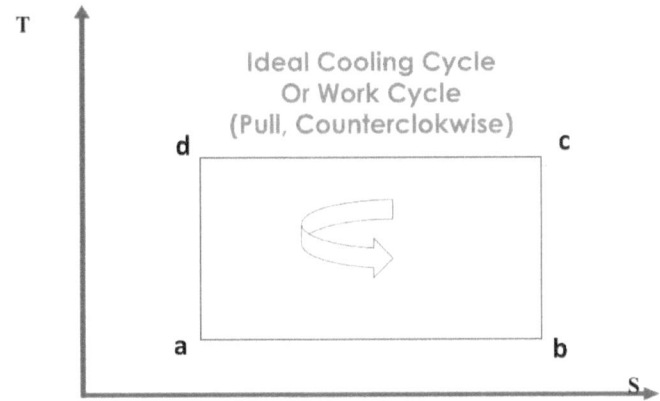

	Ideal Cooling (Carnot) Cycle	Ideal Work Cycle
a →b	Heat absorbed (Evaporation)	Resources, assets, capitalization and expenses
b → c	Work done to system (Compression)	Production and work performed
c → d	Heat rejected (Condensing)	Income and payments received
d → a	Vaporization (Expansion)	Refresh and regeneration

Figure 13 *Work Cycle*—Idealized cooling cycle may use as a metaphor of an idealized work cycle

An idealized business cycle can be constructed to describe activities in economics in figure 14:

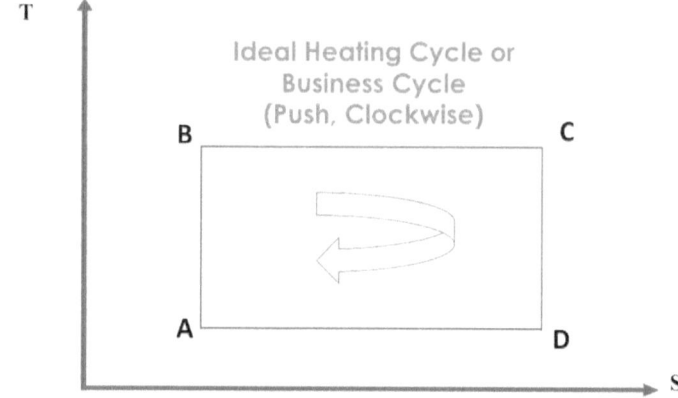

	Ideal Heating (Carnot) Cycle	Ideal Business Cycle
A →B	Compression (w/o heat added or work added)	Value added products or services
B → C	Expansion (heat added at constant temperature)	Marketing and distribution
C → D	Expansion (w/o heat added or work lost)	Products sold/consumed; service rendered
D → A	Compression (heat rejected at constant temperature)	Feedback, survey/reviews, etc.

Figure 14 *Business Cycle*—Idealized heating cycle used as a metaphor to illustrate an idealized business cycle.

In the case of production mode or supply cycle, money flows counterclockwise as that of the *cooling cycle* of refrigeration. While in the marketing mode or demand cycle, money flows clockwise as that of the *power cycle* of an engine. Figure 15 shows the demand cycle (in red) and the supply cycle (in blue) superimposed at the *equilibrium point* of the supply-and-demand diagram.

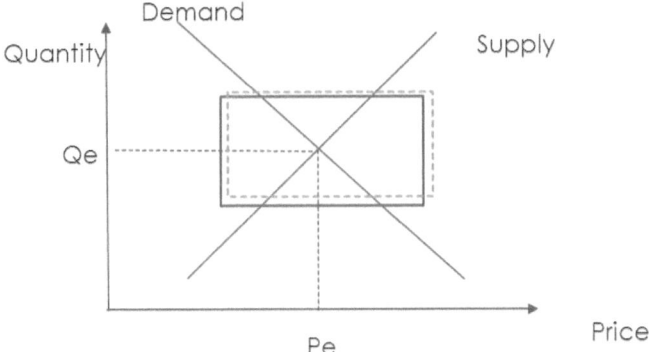

Figure 15 *Equilibrium Point*—Demand cycle (dash line) and supply cycle (solid line) TS diagrams superimposed at the equilibrium point in PQ (PV) diagram. Note: Unlike the physical system equilibrium for equation of state, this market equilibrium point is not in real time. In other word, it is a state of 'quasi-equilibrium'.

Similarly, Carnot cycles can be constructed for a production cycle, which corresponds to a cooling cycle, and a business cycle, which corresponds to a heating cycle. Production cycle examples are the production of a vineyard, an almond orchard, and a manufacturing company. Business cycle examples are grocers, supermarkets, and car distribution networks. Table 3 compares the differences between the production cycle and business cycle as well as the power cycle and the cooling cycle. Note: Cash flow at high frequency likens an engine that operates at high rotations per minute that generates more power.

Table 3 *Cycles Similitude*—Comparison of the power cycle, cooling cycle, business cycle, and production cycle. Note that goods/ services flow in a counter direction of money flow.

	Media flow	*Heat flow*	*Cycle area*	*Note*
Power cycle	Clockwise	Heat injects at high temperature and ejects at low temperature	The amount of power generated. Higher frequency implies larger power output	Work done or power generated
Cooling cycle	Counterclockwise	Heat absorbed at low temperature and ejected at high temperature	The amount of heat pumped	Energy required to pump heat from low side to high side
	Money/cash flow	*Value flow*		
Business cycle	Clockwise	Goods sold/service rendered (value diminished)	Revenue of a business cycle	Profit generated as sales completed
Production cycle	Counterclockwise	Transform raw materials into finished goods (value added)	Cost of a production cycle	Labor, cost, materials, equipment, overhead required to add value to products

Examples

Figure 16 is an example of a production cycle like a Carnot cooling cycle. Note the area representing the total production cost of 10 units. Figure 17 is an example of a business cycle like a Carnot heating cycle, and the total area is 20 units.

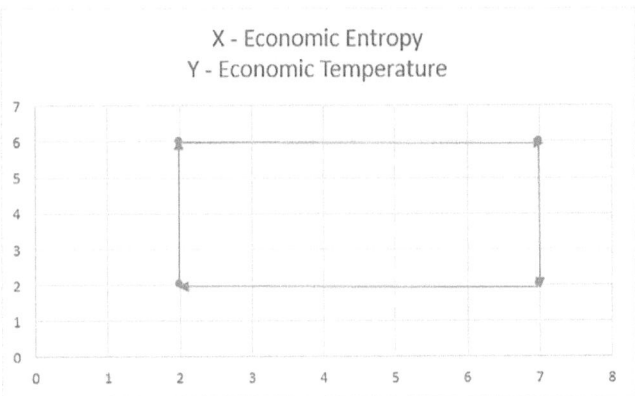

Figure 16 *Production Cycle Example*—Carnot production cycle (area = 5 × 2 = 10)

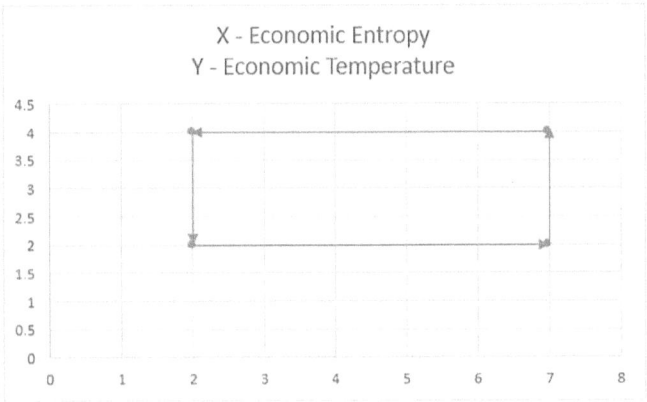

Figure 17 *Business Cycle Example*—Carnot business cycle (area = 5 × 4 = 20)

The 10-unit difference between the production cycle cost and the business cycle revenue is the net profit, which drives the economy. Like design heating and cooling equipment, the Carnot economic cycle provides visual tools to show *profit = revenue – cost*, planning and forecast.

CHAPTER 4

Macroeconomic Applications

Internet technology empowered the gig economy that the global village concept as a reality. Outsourcing and mergers are common business practices. Figure 1 shows the reasons for outsourcing: cost reduction, flexibility, fresh perspective, access to more talent, and speedy hiring process. The decision to outsource or keep in house is a challenge for all managers, planners, and operation heads. Elasticity/compressibility of products/services provides additional information for decision-makers. Following are four economic iso-processes to illustrate their significance. This chapter introduces possible macroeconomic applications via constant-price, constant-volume, constant-temperature, and constant-entropy processes.

Good Reasons for Outsource
Cost reduction
Flexibility
Access to more talent
Fresh Perspective
Speedy Hiring Process

Figure 1 *Outsourcing*—Some good reasons to outsource.

A. Constant-price process

This is the situation that goods' price remains constant *everywhere*. Suppose the living standard of Vietnam is represented by the economic temperature of T_L, and the living standard of a rich country America is maintained at the economic temperature of T_H. Figure 2 illustrates the effect of a constant price, P_H, in the PQ (PV) diagram. Figure 3 illustrates how the constant price intersects with T_L and T_H in the TS diagram. If the price remains the same, people of the high-income region have more purchasing power than people of the low-income region. People from the region of T_H travels from the United States to Vietnam with T_L will undoubtedly gain additional purchasing power represented by the shaded area in the TS diagram.

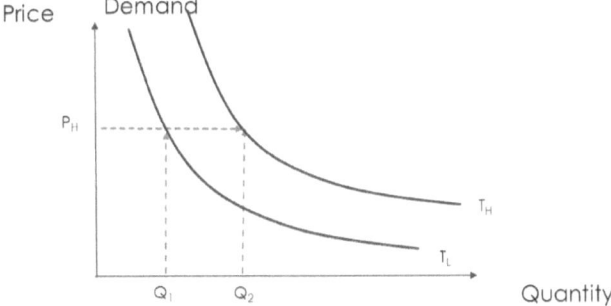

Figure 2 *Purchase Power in PV Diagram*—Here the constant price line intersects with demand curves of different living standards. It indicates that the higher income region has more purchase power in terms of quantities.

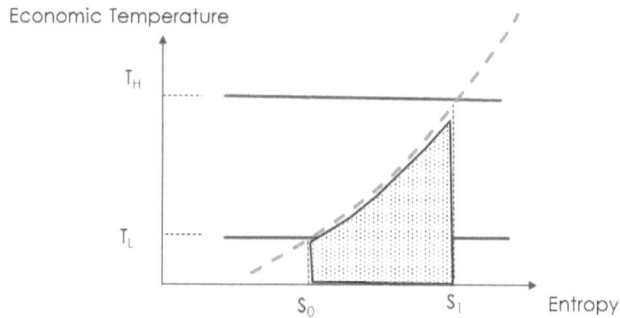

Figure 3 *Living Cost in TS Diagram*—People in higher income region are driven to low living cost region as water flow to lower ground.

On the other hand, if goods are maintained at a constant price, figure 4 illustrates that more goods, Q_2, can be produced along T_L than Q_1 produced along T_H. Consequently, the low-income region tends to draw manufacturing from the high-income region as shown in figure 5. Again, here the assumption is that goods' price is perfect elastic everywhere. *Understanding the elasticity/compressibility of products/services is essential to plan operation migration!*

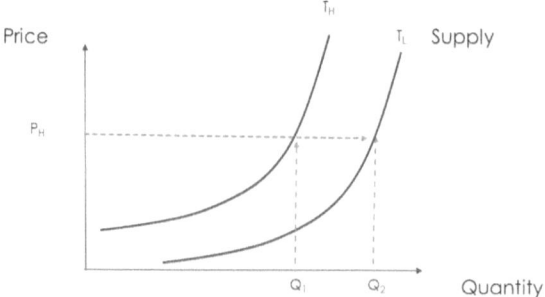

Figure 4 *Supplier Behavior in PV Diagram*—Price remains constant will cause suppliers move to the low cost region.

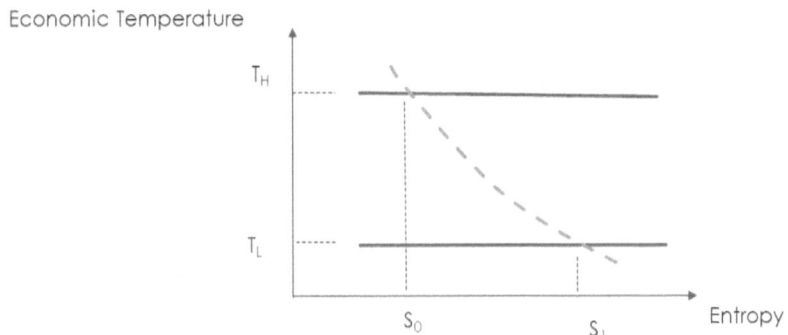

Figure 5 *Supplier Behavior in TS Diagram*—Constant-price in regions of different living costs will draw suppliers in terms of varieties.

B. Constant volume process

A good example of a constant volume process is in an auction. Item quantities are fixed at Q_0, but the price can rise sharply from P_L to P_H during the auction. Likewise, speculation can cause a hot stock price to rise quickly without increased production (see figure 6). Figure 7 illustrates the area under the isochoric line that represents the increased amount of money change hands in an auction and the speculative stock market. Auction is an inelastic market where Z increases sharply.

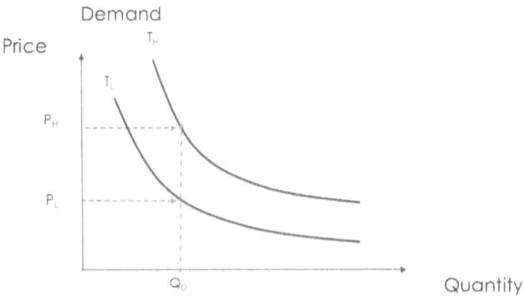

Figure 6 *Auction Psychology in PQ Diagram*—Auction with limited items is an inelastic market which drives price up sharply.

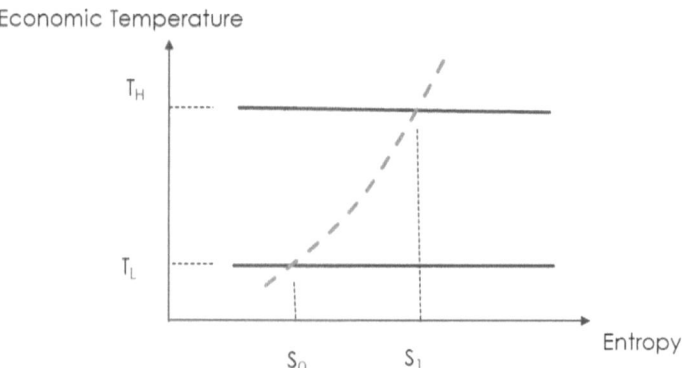

Figure 7 *Stiff Market Gain in TS Diagram*—Areas underneath the constant volume curve shows the amount of money traded increases due to higher price.

Another example is the formation of a market bubble, where a psychological effect takes place and stock price becomes inelastic/incompressible. Unfortunately, the price will eventually fall, and the bubble collapses. Figure 8 is an example of a Bulgarian stock market bubble analyzed by Eftimov (2013).

SOFIX Index

(Bulgarian Stock Exchange)

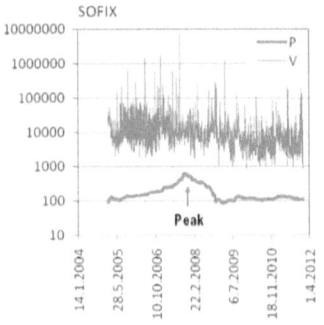
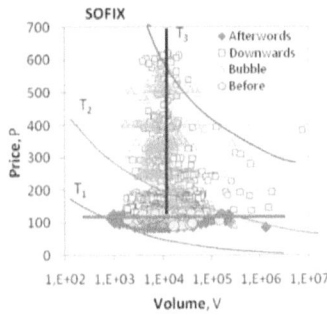

SOFIX index prior to, during and after the bubble:

a) time dependence of price and volume;

b) P-V dependence of the SOFIX index.

IDEAL GAS LAWS IN THERMOECONOMICS AND

FINANCIAL BUBBLE FORMATION

TINKO EFTIMOV 2013

Figure 8 *Market Bubble*—SOFIX Index Bubble collapsed in Bulgarian Stock Exchange (adopted from Etimov T. 2013)

C. Constant-temperature process

In the case of the isothermal process, the supply and demand curves assume the unit elasticity/compressibility. Figure 9 illustrates the isothermals in the PV/PQ diagram. Economic temperature represents the living standards of different regions. Goods produced in low labor cost and material cost will draw interests from the high living cost region and encourage trades.

Iso-thermal Process

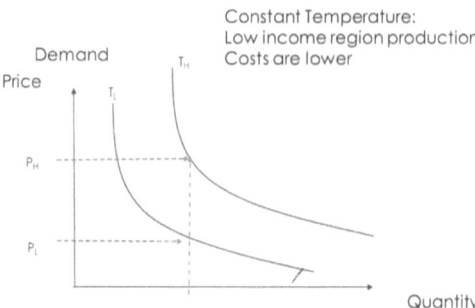

Constant Temperature:
Low income region production
Costs are lower

Figure 9 *Trading Opportunity in PV Diagram*—Demand curves of constant temperatures represent living standard of different regions that stimulates trading.

Figure 10 shows isothermals in the TS diagram. In the case of gambling, it operates along the constant temperature lines without generating profits or creating work via a Carnot engine. Therefore, from an economic point of view, gambling does not have any overall economic benefits!

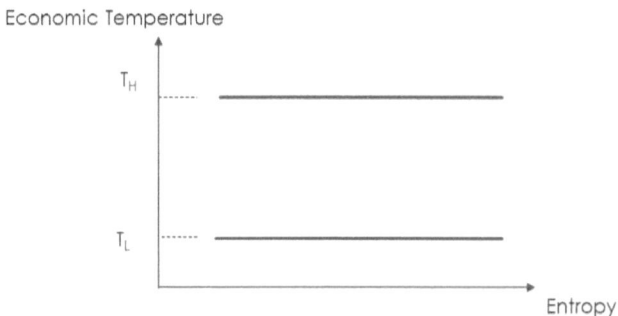

Figure 10 *Business Potential in TS Diagram*—The parallel lines indicate opportunities for outsourcing and business opportunities. Gambling tends to exchange money at the same isothermal line, e.g.: high stake gambling at luxury casino or coin gambling on the street. Therefore, no real benefits for the economy.

D. Constant-entropy process

In the cases of security, defense, emergency, rescue, political manipulations, or humanity aids, the goal of the mission transcends economic values. Military operation, donations, and voluntary services will go beyond market constraints regardless of the market constraints. Constant entropy (focus) R&D to cope with the unprecedented virus in 2020. Trump set up a "Warp Speed Project" team to concentrate human, finances, and all resources to develop a new vaccine. This situation can be called constant entropy or a focused R&D process.

From a political perspective, this is the basic difference between a centralized government and a democratic government. Capitalism advocates a high degree of freedom to improve people's living standards while communism advocates centralization and autocracy to control freedom to complete national economic policies. China's high-speed railway, satellite, and other technological developments are all based on this concept. Figures 11 and 12 illustrate such activities as an isentropic process.

This can also be a case of perfect inelastic supply caused by the following factors: firm operating close to full capacity, running out of raw materials, limited factors of production, low levels of stocks, or planning restrictions (Inelastic Supply by Pettinger, 2017, economicshelp.org).

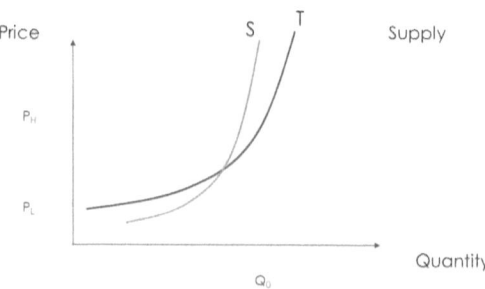

Figure 11 *Infrastructure Projects in PQ Diagram*—Many public projects may be considered as constant-entropy process of the supply curve in PQ diagram

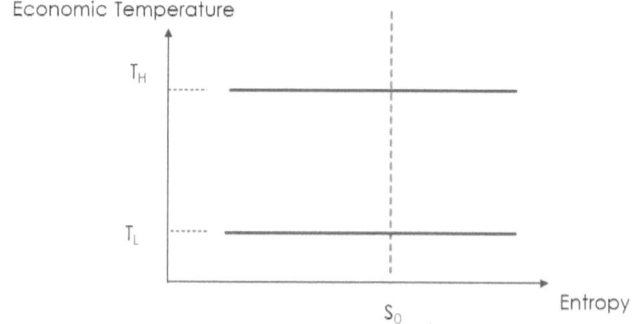

Figure 12 *Warp Speed Projects in TS Diagram*—Cost of a public project may increase or decrease due to organization structure or policy changes without involving businesses or trading.

Those four processes are extreme examples with the assumption of the market is relatively stable or near the equilibrium state. All figures are schematic and conceptual base on the equation of state of the ideal gas model. The challenges are to quantify and verify in practical cases as shown in figure 8, the Bulgarian stock exchange bubble formation example.

GDP, GNP Vectors in TS Diagram

For a country, the gross domestic product (GDP) can be calculated with the following formula:

GDP = C + G + I + NX

C = consumption or all private consumer spending within a country's economy, including durable goods (items with a lifespan greater than three years), nondurable goods (food and clothing), and services.

G = total government expenditures, including salaries of government employees, road construction/repair, public schools, and military expenditure.

I = sum of a country's investments spent on capital equipment, inventories, and housing.

NX = net exports or a country's total exports less total imports.

With a TS diagram, GDP or GNP can be charted as vectors to differentiate the kinetic energy (KE) and potential energy (PE) of the country visually. The following is an example for 2019 GDP vectors in the TS diagram (figure 13).

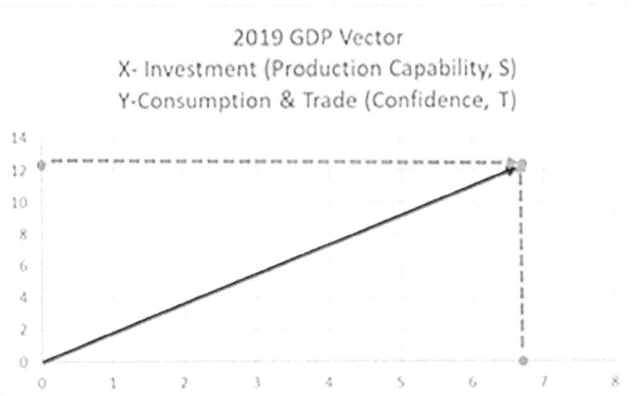

Figure 13 *GDP Vector in TS Diagram*—A GDP vector consists of two components in terms of Investment (potential energy) and Consumption & Trade (kinetic energy) of the country.

In physical science, the ideal gas model is a hypothetical construct; therefore, there is an unlimited application in the field of economics.

Effect of Quantitative Easing (QE) in Global Trade

Recently, the United States and many countries adopt the quantitative easing (QE) policy to boost the economy. Figure 13 illustrates such a concept.

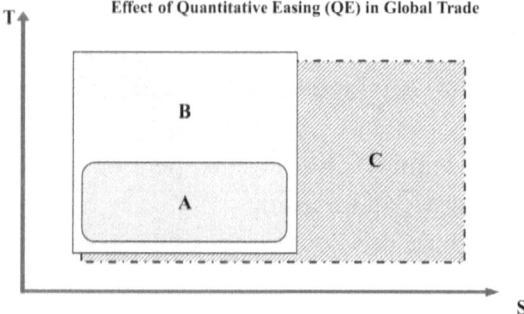

Figure 14 *Effect of QE Policy in Global Trade*—Area A represents the production cycle by a poor country. Area B represents the business cycle of a rich country. The area difference (B–A) is the profit of the rich country. When the rich country implements QE policy, it will increase the money supply and assets as indicated by area C. The rich country's profit would be (C–A).

The proposed TS diagram applications in the macroeconomic analysis are in the concept stage. Further research in this area is needed to verify its practicality and value.

Inflation

There are two types of inflation: demand-pull inflation and cost-push inflation (see figure 15). Both types of inflation can be schematically diagrammed in the TS domain as shown in figure 16 and figure 17, respectively.

Demand-pull inflation
(Inflation caused by and increase in aggregate demand)

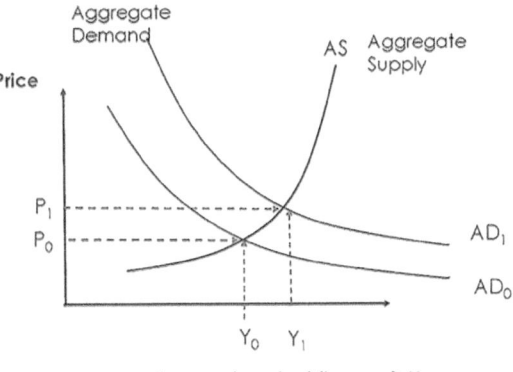

Cost-push inflation
(Inflation caused by and increase in aggregate supply cost)

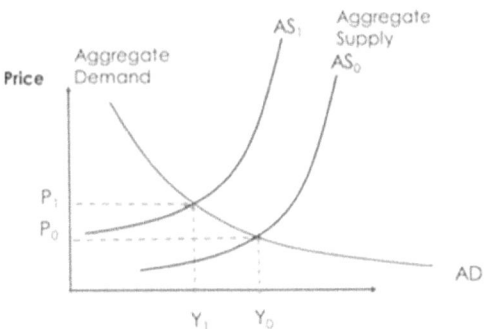

Figure 15 *Causes of Inflation*—There are two types of inflation: (a) demand-pull inflation, where AD_0 shifts to AD_1 due to higher demand while AS remains the same, and (b) cost-push inflation, where AS_0 shifts to AS_1 due to supply cost increases while the AD remains the same.

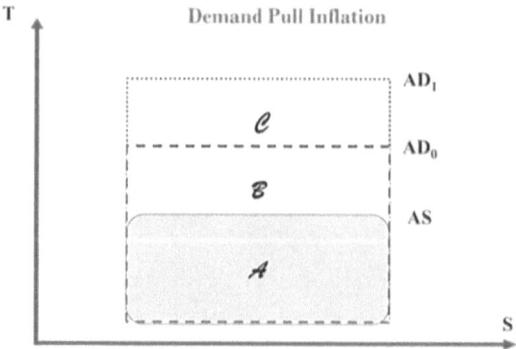

Figure 16 *Demand-Pull Inflation in TS Diagram*—Increased demand causes AD_0 shifts to AD_1 while AS remains the same. Note: Area A is the production cost, area B is the revenue at AD_0, and area C is the revenue at AD_1. In this case, there is a profit increase by C–B.

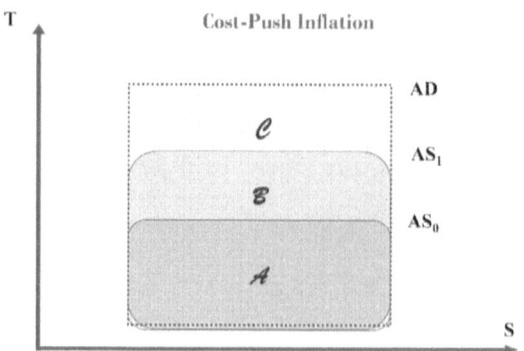

Figure 17 *Cost-Push Inflation in TS Diagram*—Increased supplier cost causes AS_0 shifts to AS_1 while AD remains the same. Note: Area A is the production cost at AS_0, area B is the production cost at AS1, and area C is the revenue at AD. In this case, there is a loss by B–A.

CHAPTER 5

Market Growth and Decline— Equation of Change

The secret things belong to the Lord our God, but the things revealed belong to us and to our children forever, that we may follow all the words of this law.

—Deuteronomy 29:29 (NIV)

A recent drought in Taiwan reminds us that rational methodology in physical science is limited (figure 1). Volti (2017, p. 13) stated the limit of rational methodology and attitude:

"All societies are faced with the problem of coping with the capriciousness of the weather. A great deal of human sufferings has been the result of the vagaries of rainfall, and history provides many examples of the tragic consequences of drought. A number of responses are possible when people are confronted with this problem. The simplest is to succumb to despair, and perhaps try to find meaning in it by attributing the drought to fate or God's will."

Figure 1 *Irrational Drought*—2021 Drought in Taiwan reminds us the limitations of rational methodology in physical science.

Weather forecasts and economic forecasts directly impact people's daily lives. Nonetheless, rational methodologies are limited in both fields. Table 1 summarizes common methodologies used for weather forecasts and economic forecasts.

Table 1 *Forecast Methodology*—Comparison of methodologies for weather forecast and economic forecast.

	Equation of state	Equation of change	Note
Weather Forecast	Ideal Gas Law	Modeling with Navier-Stokes Equation and conservation of mass and energy, and aided with simulation, statistics, etc.	Sensitive to small changes of initial condition—known as the 'Butterfly Effect'
Economic Forecast	Law of Supply and Demand (Form of Ideal Gas Law)	*Stochastic* and *heuristic* in nature. Methodologies include: mathematical optimization, statistics, game theory, econometrics, statistics, simulation, quantum mechanics (Eigen vectors, wave equation), artificial intelligence, etc.	Subjected to human factors, psychological effect, acts of God, Heisenberg's Uncertainty Principle, etc. "Phenomenological Laws" (see appendix C) may serve as the counterpart of Navier-Stokes equation to describe the spatial temporal behaviors.

Paul Samuelson and many neoclassical economists attempted to formulate the behaviors of markets in mathematical terms according to abstract laws of supply and demand. Nonetheless, there are some concerns regarding this approach:

1. The demand and supply curves discussed are hypothetical and idealized as phase space of inputs and outputs reduced in 2-D. To be useful and practical in any market, data of supply and demand should be collected as production functions of time. Therefore, the assumption of equilibrium market does not reflect the dynamics in markets such as stock market, housing market, flea market, and garage sale.

2. Supply and demand curves are independent equations of state. In other words, they are behaviors of two different

entities much like two different machines with its own characteristic performance curve. The *equilibrium* condition assumed by the intersection point of supply and demand curves is not warranted unless there exist "perfect coupling" mechanism. A practical example of "quasi-perfect coupling" is the Just-In-Time practice of Toyota Corporation supply chain logistic for production. Nonetheless, it takes strict discipline and long term planning to streamline such coupling phenomena. Therefore, mathematical optimization with abstract space is questionable.

3. Samuelson focused on mathematical analysis instead of the mechanism of the economic processes. With the proposition of gas model for a market, it is reasonable to explore the feasibility of applying methodology used to analyze transport phenomena in the field of economics. Common transport phenomena in physics consist of energy transfer, momentum transfer, and mass transfer via Fourier's Law, Newton Fluid Law, and Fick's Law, respectively. The similarities among these laws allude us to incorporate economic processes (e.g., currency flow, goods flow, information flow, technology diffusion, etc.). Phenomenological Laws (see appendix C) presents such opportunity with empirical data analysis.

Economic Growth and Equation of Change (EOC)

Mimkes (2006) formulated an elegant two-dimensional calculus for economic systems. With this calculus framework, he developed three laws for economics that correspond to the three laws of thermodynamics (see appendix A). Table 2 is a summary of related terminologies. Though the calculus is mathematically sound and elegant, formulating putty function is a theoretical challenge. However, the economic system is a function of time beyond a two-dimensional description. Furthermore, the economic system involves large sets of variables—tangible as well as intangible. In physical science, such systems are described in phenomenological laws and related frameworks.

Table 2 *Predictablilty*—Terminologies used in the development of two-dimensional calculus for economic systems according to Mimkes (2006).

Terminologies	Predictable	Non-predictable	Note
Physics	*Ex ante*	*Ex post*	
Economics	Clay function	Putty function	Edmund Phelps (1963) Leif Johnsen (1959)
Calculus	Exact differential	Inexact differential	

Prigogine (1984) summarized the equation of changes for numerous physical variables in terms of phenomenological laws (see appendix C). The term *phenomenological* implies that scientists do not fully understand the molecular level mechanism or logic structure associated with a particular phenomenon. Many transport phenomena such as heat conduction, diffusion, or fluid flow are governed by the phenomenological laws, but we have yet limited understanding.

Phenomenological laws:
$$\mathbf{J} = \mathbf{L}\,\mathbf{X}$$
{Rate of changes} = {coefficients matrix} × {driving forces}

where $\mathbf{J}$ is a vector formed by $J1$, $J2$, $J3$,...J_n; $\mathbf{X}$ is a vector of $X_1, X_2, X_3,...X_n$; and $\mathbf{L}$ is a n by n matrix with elements $L_{11}, L_{12},...L_{nn}$.

This is even more true in the biological world and medical fields. The contemporary COVID-19 is a prominent example of limited understanding how virus is spread in our environment. In other words, we cannot formulate a set of EOC for the virus spread exactly. In conclusion, the attempt to develop a set of economic equations of change to model economic processes as the counterpart of physical science has limited success because of the psychological effects and external factors. Conceptually, the economic equation of state (EEOS) and economic equation of change (EEOC) may be related as shown in figure 2. From State A transition to State B, there are infinite number of processes. Figure 2 depicts only two paths: typical growth S-curve (top) and push-pull oscillation (bottom). The S-curve on the top is the general charteristic of all technology development that may be used to depict EEOC. The technology development or growth S-curve can be classified into three stages: Stage One—Incubation/Launching; Stage Two—Growth, Stage Three—Mature and Market Saturation. At any point of Stage Three, there may a 'bifurcation point' where the growth either continue, stop, or even decline. Prigrogine theorized this is due to 'dissipative structures'. In social science, this is an interesting topic of many scholars, such as Jim Collins who authored the book: "From Good to Great". Figures 4, 5 and 6 summaize characteristics and strategies for each stage, respectively. S-curve is the total number of users versus time, or an arbitrary unit, such as as volume, or market size to reflect the economic temperature for a particulr product, service, or technology.)

States and Processes via Equation of Change (EOC)
(Equilibrium vs Non-Equilibrium)

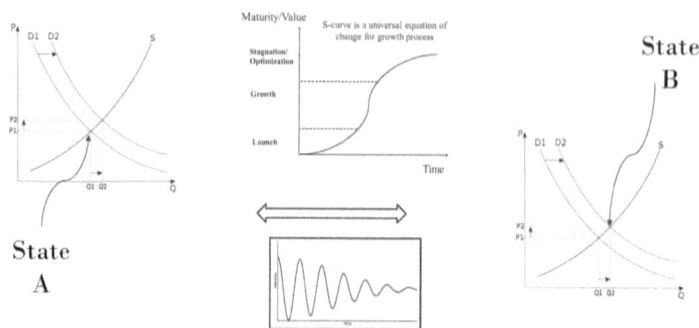

Figure 2 *States & Processes*—Equilibrium State A is subjected various possible processes to arrive a new equilibrium state B via economic equation of change (EEOC).

Stages of S-Curve

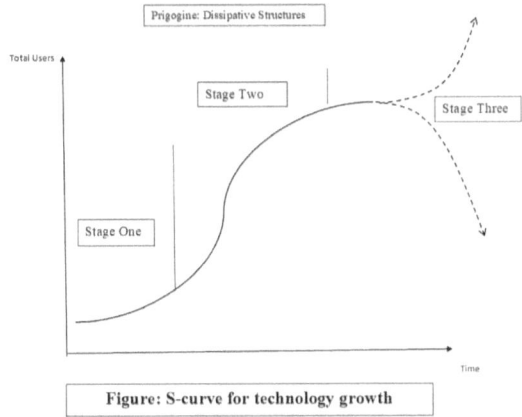

Figure: S-curve for technology growth

Figure 3 *Growth Curve*—S-curve is a universal equation of change for growth process. Generally it is classified into three stages: 1. rapid growth, 2. steady growth and 3 slowed growth. Bifurcation may take places pending governing mechanism either continue to grow or decline.

Stage One

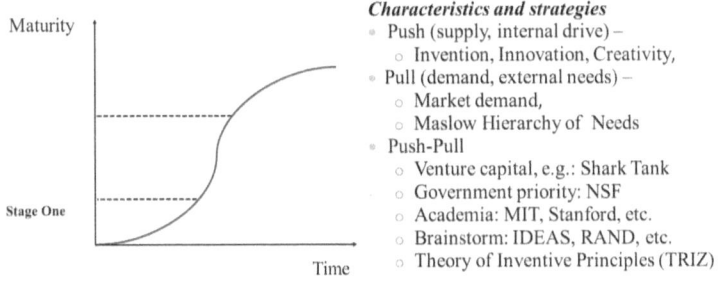

Characteristics and strategies
- Push (supply, internal drive) –
  - Invention, Innovation, Creativity,
- Pull (demand, external needs) –
  - Market demand,
  - Maslow Hierarchy of Needs
- Push-Pull
  - Venture capital, e.g.: Shark Tank
  - Government priority: NSF
  - Academia: MIT, Stanford, etc.
  - Brainstorm: IDEAS, RAND, etc.
  - Theory of Inventive Principles (TRIZ)

Figure 4 *Stage One*—It characterized by constructivism, high in uncertainty, degree of freedom, potential energy, or economic entropy. Note: high uncertainty in this stage is referring to external factors instead of the entropy of the system.

Stage Two

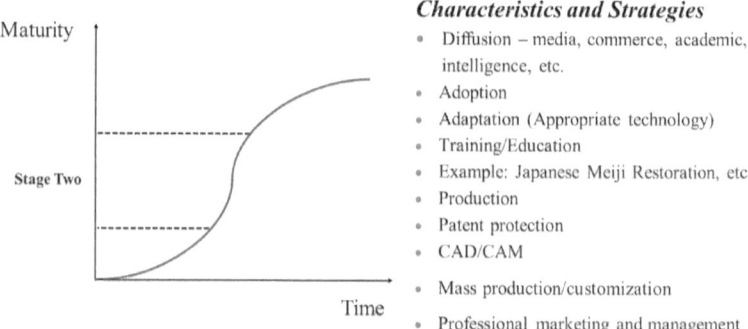

Figure 5 *Stage Two*—It characterized by increased communication.

Stage Three

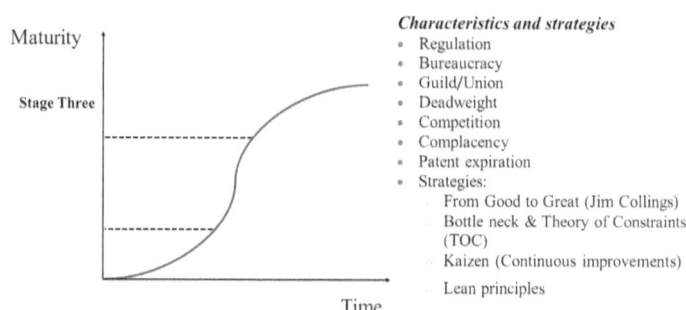

Figure 6 *Stage Three*—It characterized by determinism, low degree of freedom and potential.

A practical manifestation of the *S*-curve is the Morningstar Style Box used to capture the characteristics of a mutual fund or ETF investment as shown in figure 7. The box uses the three stages of an inversely transposed *S*-curve to characterize the investment funds. Please note increased risk at the early growth stage of the *S*-curve.

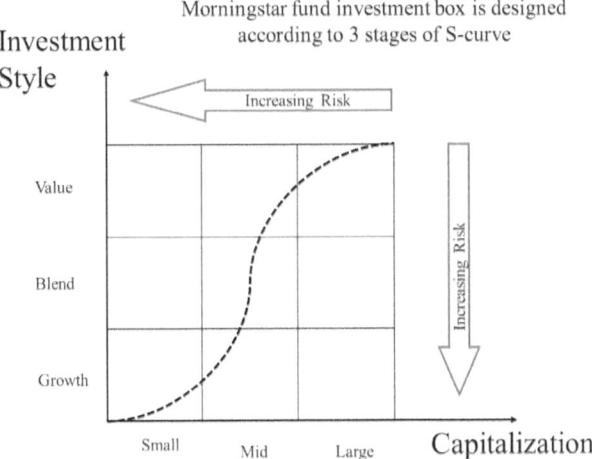

Figure 7 *Mutual fund Investment Box*—Also known as the Morningstar style box captures the three stages of a S-curve in terms of growth and capitalization. Note increased risk at the early growth stage of an S-curve.

Another well-known example of the equation of change is the Moore law for semiconductor technology as shown in figure 8. It is a prediction made by an American engineer Gordon Moore in 1965 that the number of transistors per silicon chip doubles every year.

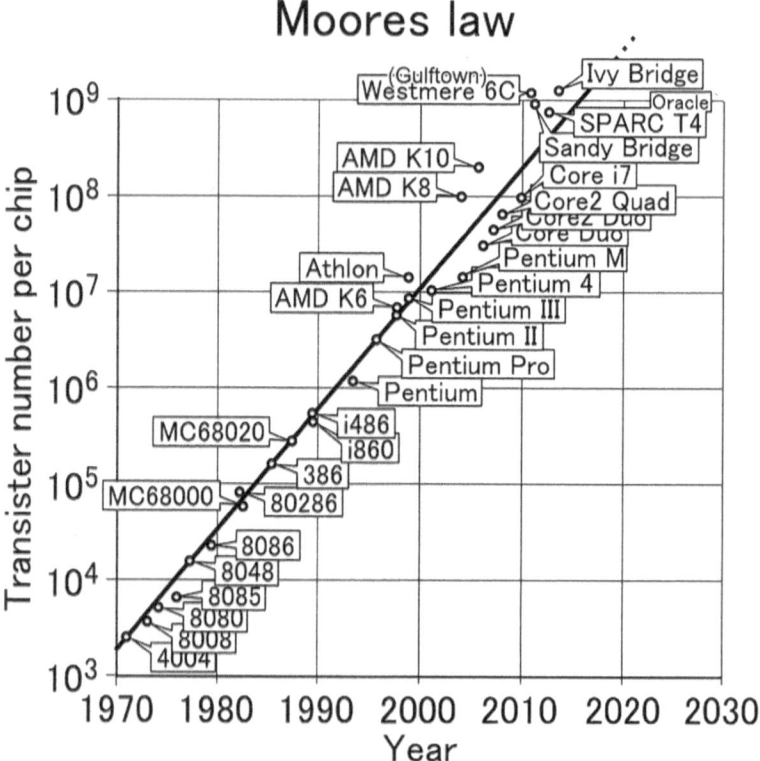

Figure 8 *Moore's Law*—It predicts the number of transistors doubles every year as an Equation of Change of the semi-conductor technology. Consequently, it affects prices of related products.

In thermodynamics, Boltzmann worked out the entropy of gas of microstates as:

$$S = ln(w) \qquad (1)$$

S is the logarithmic function of the total number of the microstates, *w*. In other words, entropy is proportionally to the number of options, possibilities, or degree of freedom. Therefore, entropy is a measure of uncertainty.

According to the second law of thermodynamics, the entropy of the universe increases over time, i.e.,

$$\frac{dS}{dt} > 0 \quad (2)$$

The simplest example of entropy increase is the population increase over time (figure 9).

The evolution of the world population

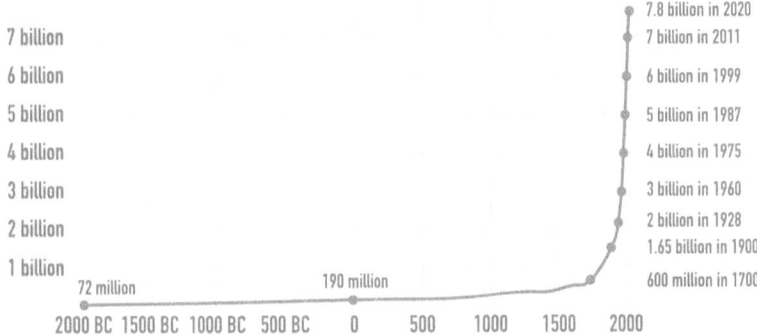

Figure 9 *Population Growth*—Population growth is an example of entropy increases over time. Economics grows with population growth.

Claude (1948) related entropy to information and stated, "Information is the negative of entropy." This means that, for a system, an increase in information is equal to a reduction in entropy, or

$$\Delta I = -\Delta S \quad (3)$$

Here, I symbolizes the amount of information of a system while S symbolizes the entropy (uncertainty, options, or degree of freedom). Please note that I refers to usable information or more precisely worthy intelligence. In plain language, information entropy S is the amount of noises.

Scholars and scientists have claimed that equations (1)–(3) are universally valid, both in physical science and in social science. In this case, we can use an example to illustrate the concept: The information contained in the strands of DNA can be used by a living cell to make a copy of itself; the entropy of the materials to make the new cell has been reduced. Another example is that technology advancement offers people more goods and options to buy. The more information (intelligence), I, about a product we have, the less uncertain (S) about our decision to buy or not to buy that product (see figure 10).

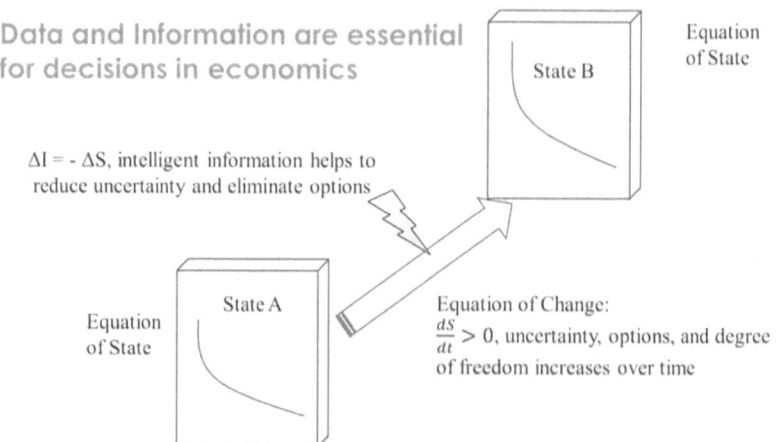

Figure 10 *Data vs. Information*—The economic equation of change for entropy increases over time. To reduce the uncertainty in economic activities, data and information are essential in decision-making.

Price Determinations

Microscopically, an original equipment manufacturer (OEM) can set the price according to a price model, such as the following:

Total price = labor cost + material cost + overhead + profit (4)

Macroscopically, price is usually determined dynamically according to the supply curve and demand curve. Figure 11 is a hypothetical example of a housing market.

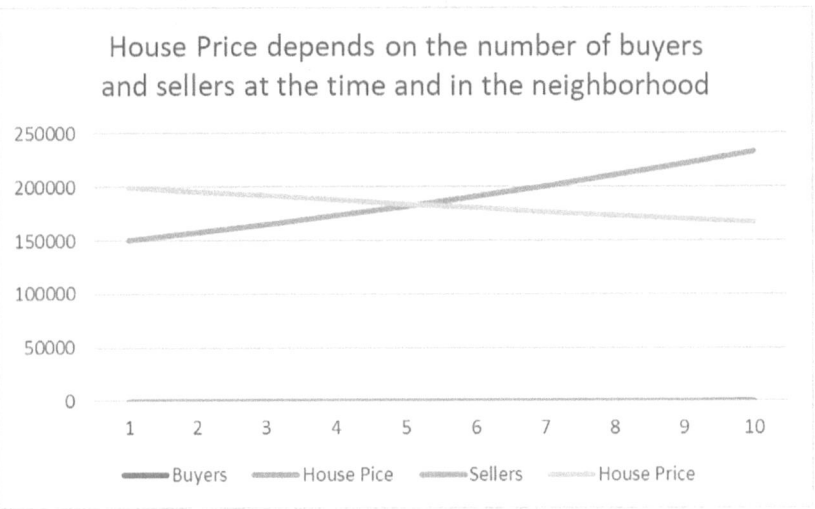

House Price depends on the number of buyers and sellers at the time and in the neighborhood

Figure 11 *Housing Market*—Price in a housing market is determined by the numbers of buyers and sellers.

Price may also be determined by the level of interest or the economic temperature (T), such as items in an auction and a garage sale. Real estate property appreciation, used car depreciations, and currency inflation are other examples of prices that fluctuate over time. Therefore, economic temperature is a measure of value rather than price. Essentially price is more of a tag and Price ≠ Value.

Forecast

Equations of state presented in chapters 2 and 3 are based on rational methodology. As for the equation of change, time must be taken into consideration. In physical science, equations of changes have been studied in solid mechanics, fluid mechanics, heat transfer, and mass transfer as a family of transport phenomena. In physical chemistry, such phenomena are characterized as an *irreversible pro-*

cess, and the system entropy increases over time. (Refer to appendix C, Phenomenological Laws).

Figure 12 shows the relationship of the economic equation of state (EEOS) and the economic equation of change (EEOC). The EEOC shares the uncertainty as in quantum physics and weather forecast. Nonetheless, recent years' successes in a search engine by applying some quantum tools, such as eigenvector and eigenvalue to guide the changes, are intriguing and inspiring. On the other hand, neural networks, big data, and artificial intelligence may also be employed to guide the EEOC along with Internet data analytics. Nonetheless, such rational techniques are yet to be developed and validated.

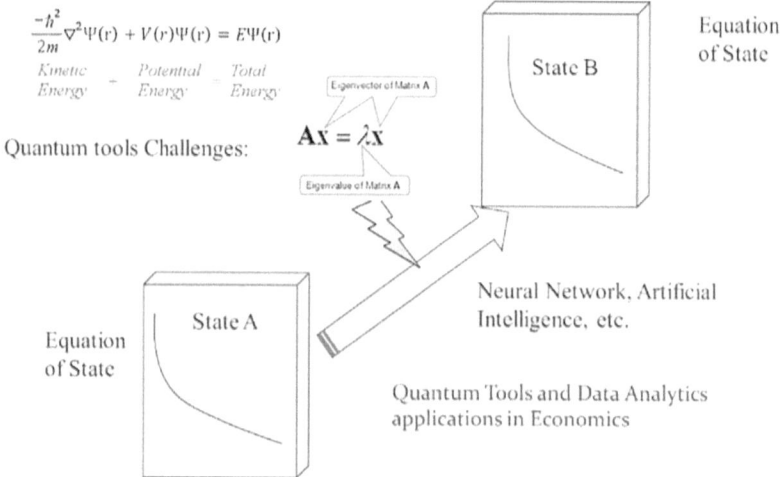

Figure 12 *Intelligent Tools*—Economic equation of state is only a snapshot of the economic activities. Modern tools may be applied to guide the economic equation of change of the economic processes.

Acts of God

Volti (2017) indicated that rational attitude focused on abstract logic and isolated from reality. The economic system is not only governed by the laws of supply and demand but also by laws of higher order such as psychology, politics, culture, and even divine intervention such as the drought in Taiwan alluded in figure 1 and depicted in figure 13.

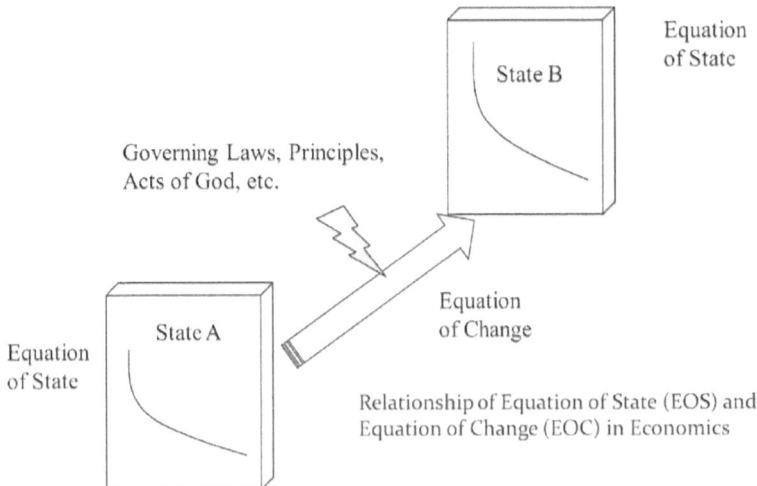

State B

Equation of State

Governing Laws, Principles, Acts of God, etc.

State A

Equation of State

Equation of Change

Relationship of Equation of State (EOS) and Equation of Change (EOC) in Economics

Figure 13 *Risk Management*—Economic is subject to laws of higher order (e.g., psychology, politics, culture, and even divine interventions).

Joseph's story in chapters 37–41 of the Book of Genesis is a famous case of Divine Intervention:

> When two full years had passed, Pharaoh had a dream: He was standing by the Nile, when out of the river there came up seven cows, sleek and fat, and they grazed among the reeds. After them, seven other cows, ugly and gaunt, came up out of the Nile and stood beside those on the riverbank. And the cows that were ugly and gaunt ate up the seven sleek, fat cows. Then Pharaoh woke up. He fell asleep again and had a second dream: Seven heads of grain, healthy and good, were growing on a single stalk. After them, seven other heads of grain sprouted—thin and scorched by the east wind. The thin heads of grain swallowed up the seven healthy, full heads. Then Pharaoh woke up; it had been a dream. (Genesis 41:1–7)

Then Joseph said to Pharaoh, "The dreams of Pharaoh are one and the same. God has revealed to Pharaoh what he is about to do. The seven good cows are seven years, and the seven good heads of grain are seven years; it is one and the same dream. The seven lean, ugly cows that came up afterward are seven years, and so are the seven worthless heads of grain scorched by the east wind: They are seven years of famine. (Genesis 41:24–26)

The story went on that Joseph advised Pharaoh:

And now let Pharaoh look for a discerning and wise man and put him in charge of the land of Egypt. Let Pharaoh appoint commissioners over the land to take a fifth of the harvest of Egypt during the seven years of abundance. They should collect all the food of these good years that are coming and store up the grain under the authority of Pharaoh, to be kept in the cities for food. This food should be held in reserve for the country, to be used during the seven years of famine that will come upon Egypt, so that the country may not be ruined by the famine. (Genesis 41:34–36)

This story highlights the economic changes subject to higher laws and divine power beyond rationality. On the other hand, this is a case illustrating that intelligent information reduces uncertainty or entropy ($\Delta I = - \Delta S$): Joseph had Divine Intelligence to render a timely decision to mitigate the famine, and he was immediately appointed as the commissioner in charge of Egypt.

Even in the case of lifeless things that make sounds, such as the pipe or harp, how will anyone know what tune is being played unless there is a distinction in the notes? (1 Corinthians 14:7)

CHAPTER 6

Market Is a Garden

Now the LORD God had planted a garden in the east, in
Eden; and there He put the man He had formed.

—Genesis 2:8

In chapter 1, Hanauer (2019) uses a garden as the metaphor for the
market. Human history started in the Garden of Eden according to the
Book of Genesis. God blesses us with talents, gifts, and skills to manage
natural resources and to enjoy abundant lives. Nonetheless, Adam and
Eve failed to respect God's command: "Do not eat the fruits of the tree
of knowledge of good and evil. God did not curse Adam, but instead,

> Cursed is the ground because of you;
> through painful toil you will eat food from it all
> the days of your life. It will produce thorns and
> thistles for you, and you will eat the plants of the
> field. By the sweat of your brow you will eat your
> food until you return to the ground, since from it
> you were taken; for dust you are and to dust you
> will return. (Genesis 3:17–19)

Therefore, all people must "toil to eat" or "to work to make a
living." Nonetheless, the ground is cursed with "thorns and thistles."

Ginsburg (1975) stated a theorem as a parody of the laws of thermodynamics in terms of a person playing a game that describes the nature of the downfall of mankind:

0. *There is a game. (a consequence of the zeroth law of thermodynamics)*
1. *You can't win. (a consequence of the first law of thermodynamics)*
2. *You can't break even. (a consequence of the second law of thermodynamics)*
3. *You can't even get out of the game. (a consequence of the third law of thermodynamics)*

Consequently, people must face the challenges to understand the game and formulate game strategies. Nonetheless, God did not abandon us, and He has outlined numerous principles about economics, nature of money system, and wealth management in addition to natural resources and daily provisions. The following are some examples according to the living water economics:

Money as a medium

Monetary system has gone through the use of gold, silver, paper money, check, credit card, Apple Pay, PayPal, bitcoin, and blockchain. The properties of ideal money are as follows: durable, portable, divisible, stable, etc. Whatever the form of the money we use, it is merely a certificate of a credit for the person who uses it. Think about it. A blank check has no value until someone puts a signature on it. If the signature were the President of the United States, it would be indeed a valuable piece of paper. According to living water economics, money is a medium that is meant to circulate in economic cycles and be replenished.

Jesus taught us about the nature of the money in Matthew 6:19–21:

> Do not store up for yourselves treasures
> on earth, where moths and vermin destroy, and

where thieves break in and steal. But store up for yourselves treasures in heaven, where moths and vermin do not destroy, and where thieves do not break in and steal. For where your treasure is, there your heart will be also.

Earthly treasures are subject to losses and destruction. The value of treasure lies in people's hearts. On an occasion, Jesus praised a widow's offering in Matthew 12:42–43:

But a poor widow came and put in two very small copper coins, worth only a few cents. Calling His disciples to Him, Jesus said, "Truly I tell you, this poor widow has put more into the treasury than all the others."

Furthermore, in Matthew 22, we read that a group of Pharisees confronted Jesus and asked for a yes or no answer on whether it was lawful to pay taxes to Caesar, the Roman emperor. Jesus responded by asking whose face was engraved on the coins used to pay the taxes. When they answered it was Caesar's face, Jesus replied with the now-famous verse, "Then render to Caesar the things that are Caesar's; and to God the things that are God's" (Matthew 22:21 NASB). This famous verse clearly pointed out that the value of earthly money is in the government system that issues it as a token of the degree of the trust. Thus, economics in the world really is a matter of a system of trust. Whenever the trust is degraded or breached, the economic system collapses.

Wealth growth via economic cycles

Matthew 6:24 reminds us that money cannot be or should not be our master:

No one can serve two masters. Either you will hate the one and love the other, or you will be devoted to the one and despise the other. You cannot serve both God and money.

Therefore, we are urged to be the master of money through wise management. This principle is given in the three servants' parable (Matthew 25:14–27).

Parable of the three servants

Again, the Kingdom of Heaven can be illustrated by the story of a man going on a long trip. He called together his servants and entrusted his money to them while he was gone. He gave five bags of silver to one, two bags of silver to another, and one bag of silver to the last—dividing it in proportion to their abilities. He then left on His trip.

The servant who received the five bags of silver began to invest the money and earned five more. The servant with two bags of silver also went to work and earned two more. But the servant who received the one bag of silver dug a hole in the ground and hid the master's money.

After a long time their master returned from his trip and called them to give an account of how they had used his money. The servant to whom he had entrusted the five bags of silver came forward with five more and said, "Master, you gave me five bags of silver to invest, and I have earned five more."

The master was full of praise. "Well done, my good and faithful servant. You have been faithful in handling this small amount, so now I will give you many more responsibilities. Let's celebrate together!"

The servant who had received the two bags of silver came forward and said, "Master, you gave me two bags of silver to invest, and I have earned two more."

The master said, "Well done, my good and faithful servant. You have been faithful in handling this small amount, so now I will give you many more responsibilities. Let's celebrate together!"

Then the servant with the one bag of silver came and said, "Master, I knew you were a harsh man, harvesting crops you didn't plant and gathering crops you didn't cultivate. I was afraid I would lose your money, so I hid it in the earth. Look, here is your money back."

But the master replied, "You wicked and lazy servant! If you knew I harvested crops I didn't plant and gathered crops I didn't cultivate, why didn't you deposit my money in the bank? At least I could have gotten some interest on it."

This is a clear illustration of how money should be used by putting into the earthly economic system to generate wealth. In other words, money should be cycled, flowed, and not to be hoarded. According to living water economics, wealth is generated via the combination of a cooling cycle of production and a heating or power cycle of marketing and sales.

Retaining Heart illustration in Sower's Parable

Like the garden's metaphor for the market, Jesus's parable of the sower can be a metaphor to explain the uses of the TS diagram in economics. First, economic entropy can be construed as the potency of the ground while the types of hearts are a metaphor of the economic temperature in degree of interest and trust. Jesus told sower's parable in Matthew 13:3–9. The parable alluded heart conditions in response of the gospel message. The parable may also apply in the case a business endeavor. The four heart conditions are summarized in Table 1. Production of a retaining heart is further illustrated in a TS diagram as shown in Figure 1.

Table 1 *Sower's Parable*—Jesus told a sower parable that is also true in the context of business endeavor.

Parable	Heart Condition
A farmer went out to sow his seed	Message, vision, or ideas
As he was scattering the seed, some fell along the path, and the birds came and ate it up.	Ignorant Heart—failed to see the value of ideas and catch the opportunity.
Some fell on rocky places, where it did not have much soil. It sprang up quickly, because the soil was shallow. But when the sun came up, the plants were scorched, and they withered because they had no root.	Rocky Heart—enthusiastic at the ideas first but lack determination to follow through.
Other seed fell among thorns, which grew up and choked the plants.	Unfocused Heart—easily distracted by other ideas and commitment to endure challenges.
Still other seed fell on good soil, where it produced a crop—a hundred, sixty or thirty times what was sown.	Retaining Heart—committed and dedicated. Produce 30 times, 60 times, or 100 times according resources availability

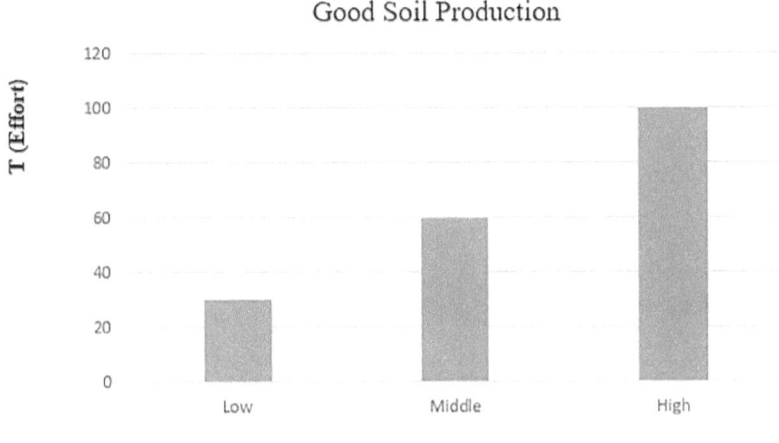

Good Soil Production

Figure 1 *Productivity and Heart Condition*—Crop production of a good soil or retaining heart are illustrated in a TS diagram.

Hanauer (2019) stated that price does not equate to value. In economics, the economic temperature has a connotation of value instead of price. The quality of people and disciplined practices are essential for successful production and business.

> See, I am doing a new thing! Now it springs up; do you not perceive it? I am making a way in the wilderness and streams in the wasteland. (Isaiah 43:19)

CHAPTER 7

History Is a River of Living Water

Then he showed me a river of the water of life, clear as crystal,
coming from the throne of God and of the Lamb, in the middle
of its street. On either side of the river was the tree of life,
bearing twelve kinds of fruit, yielding its fruit every month;
and the leaves of the tree were for the healing of the nations.

—Revelation 22:1–2

Since the first Industrial Revolution, living water flows in history as a
river of life. Regardless the advancement of technology and sciences,
air and water are essential in people's lives today as it was in the day
of the Garden of Eden.

The ideal gas concept was inspired by the use of steam to power
the steam engine. It facilitated the development of thermodynamics
in physical science. It also inspired the development of the economics of equation of state (EEOS) and product compressibility factor,
Z, as in physical science.

The EEOS is a prototype for the supply and demand curves.
The compressibility factor is an index for price *negotiation effort*. It
would be a valuable endeavor to establish libraries of various products and services (e.g., agricultural products, electronic products,
industrial products, and consumer products).

Unlike physical science, economics must include *people* as Hanauer (2019) pointed out while the rationality methodology tends to isolate people from the discussion to exclude psychological effects. Introducing the TS diagram in economics as a tool for market fundamental analysis opens the option to include people and psychological effects in discussions. It enables the discussions of intangible factors, such as psychology, intelligence, belief, faith, and spiritual matters beyond rationality.

Metaphorically, economic temperature (T) represents the kinetic energy (KE) or psychological effect of an economic entity while economic entropy (S) represents the potential energy (PE) or the production capability of the economic entity. All contemporary data analysis techniques, such as statistics, survey, sampling, rating, voting, data analytics, could be applied and not to isolate people from the study.

The economic iso-processes in the PV and TS diagrams provide a set of qualitative tools for governments and institutions to explore regional economy planning.

The economic equation of change (EEOC) is far above and beyond mathematical modeling. However, we may apply neural networks, artificial intelligence, big data, and contemporary searching techniques of eigenvectors for limited forecasting.

Insights gained from the rational model of the economic ideal gas model are not in conflict with the biblical teaching regarding the characteristics of the money system. In fact, it endorses the scripture's principles about the money system, namely, money is merely a token empowered by faith, credit, and trust; cash flow and wealth growth; information; and intelligence. Economic effectiveness must discern environmental factors and the quality of people.

Information Versus Intelligence

He said, "Go and tell this people: 'Be ever hearing, but never understanding; be ever seeing, but never perceiving.'"

—Isaiah 6:9

When Shannon (1948) formulated the information entropy, he meant that entropy is a measure of noises. However, the Internet is inundated with too much information and becomes noise. The decision that people need to make is really *intelligence*—credible and valuable information. Weather forecasting today employs sophisticated instruments and modeling. Likewise, forecast in economics requires modeling and intelligent data collection. All these demand people with critical thinking and discipline.

Genesis of Economics

In the Book of Genesis 2:7:

> Then the Lord God formed a man from the dust of the ground and breathed into his nostrils the breath of life, and the man became a living being.

The original Hebrew word for *breath* is נְשָׁמָה (*nesh-aw-maw*). It means blast, spirit, life, etc. Regardless of the rich exposition of the word *breath* in Hebrew, God's act in this verse is expressed as a flow of air physically. The adoption of a gas model to describe economic activities is consistent with the fact that *breathing* is the designated physical mechanism to sustain human life at large. Prior to *The Wealth of Nations* book, Adam Smith coined the term *the invisible hand* (see table 1):

> The Invisible Hand is an economic concept that was first introduced by Adam Smith in The Theory of Moral Sentiments, written in 1759. The Invisible Hand is a metaphor describing the unintended greater social benefits and public good brought about by individuals acting in their own self-interests though according to Smith it was literally Divine Providence, that is, the hand of God working to make this happen. (Invisible Hand, wikipedia.org)

Table 1 *Divine Inspiration*—Breathing and spirit are unique to human life and economic activities. The gas model is a manifestation of the invisible hand coined by Adam Smith.

Hebrew	רוח (rauch)	נְשָׁמָה (neshamah)
English	Breath of Life, wind	Spirit
Human	Yes	Yes
Animals & Living creatures	Yes	No
And the LORD God formed man of the dust of the ground, and breathed into his nostrils the breath of life; and man became a living soul. (Genesis 2:7)		
And with that he breathed on them and said, "Receive the Holy Spirit. (John 20:22)		

The thermodynamics of ideal gas provides a viable tool to describe the *invisible hand* qualitatively as well as quantitatively (see figure 1).

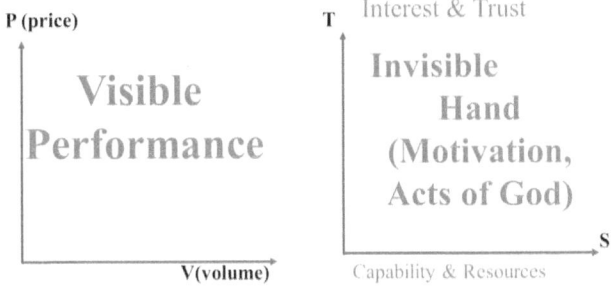

Figure 1 *Duality*—PV and TS diagram used in thermodynamics describing the duality of gas properties is a set of valuable tools to aid the illustrations of economic systems metaphorically, qualitatively, and quantitatively.

The 2023 bank crisis, the 2008 subprime mortgage crisis, the Great Depression (1929–1939), the Russian-Ukraine war, and other adverse events reveal the limitations of current economic theories that ignore God's role in financial matters. The motto "In God We Trust" on US currency reminds us to honor God's grace and blessings (see Figure 2). Jesus said, "So give back to Caesar what is Caesar's, and to God what is God's." Therefore, the motto calls for God's protection of the country in times of crisis.

Figure 2. In God We Trust—The motto on US currency was approved by President Eisenhower as a joint resolution on July 30, 1956.

The concept of the motto dates back to the American Revolution (1748–1785), when both British and American people sought divine protection. The motto was inspired by Psalm 115:9–11: "…You who fear him, trust in the LORD—he is their help and shield," and was later invoked by President Lincoln during the Civil War and by President Eisenhower during the conflict with communist Russia for divine protection. Nonetheless, the invisible hand blessed the US with the strongest economic growth in the world during that period.

As we enter the digital age, people are predisposed to using credit cards and digital money. John Nash described 'ideal money' in 2002. Satoshi Nakamoto launched cryptocurrency in 2009, which later evolved into blockchain technology. Does such shift imply the American are moving away from the divine protection of "In God We Trust". I wonder why the US currency is losing its value?

For protection and blessings, let us seek the Lord of the invisible hand—the source of Living Water, as Jesus said: 'He who believes in Me, as the Scripture said, "From his innermost being will flow rivers of living water"' (John 7:38)."

Give, and it will be given to you. A good measure, pressed down, shaken together and running over, will be poured into your lap. For with the measure you use, it will be measured to you. (Luke 6:38)

APPENDIX A

Thermodynamic Laws

As the rain and the snow come down from heaven, and do
not return to it without watering the earth and making it
bud and flourish, so that it yields seed for the sower and
bread for the eater, so is my word that goes out from my
mouth: It will not return to me empty, but will accomplish
what I desire and achieve the purpose for which I sent it.

—Isaiah 55:10–11

1970 Nobel Economics laureate Samuelson (1970) acknowledged
that the maximum principles in analytical economics were due to
the influence of Gibbs-like thermodynamics. Willard Gibbs (1874–
1878) is known for his work to graphically represent the thermodynamic properties, namely: energy, volume and entropy.

Water is an essential substance that powered our livelihood and
economics. Following Gibbs work, we can visualize the marvelous
properties of water in 3-Dimensional topological rendering (see
Figure 1).

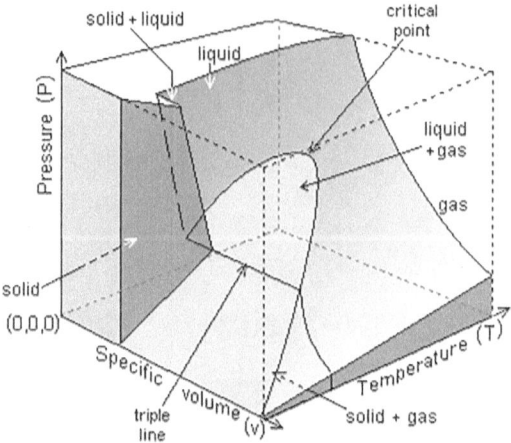

Figure 1 *Water in a 3D Model*—Phase Surface of water in temperature-volume-pressure coordinates. (modern rendering of Gibbs' thermodynamic graphical method). (Faghri and Zhang 2006)

Samuelson described Gibbs as "Yale's great physicist." Gibbs strongly influenced the education of the economist Irving Fisher who was awarded the first Yale PhD in economics in 1891. Gibbs's student Edwin Bidwell Wilson passed the Gibbsian knowledge to Samuelson. Gibbs elaborated on the first and second laws of thermodynamics and expanded the envelope of applications in fields beyond mechanical systems to include chemical physics and economics. As a reference, the three laws of thermodynamics are summarized below:

Laws of thermodynamics

- First law of thermodynamics: When energy passes, as work, as heat, or with matter, into or out from a system, the system's internal energy changes in accord with the law of conservation of energy.
- Second law of thermodynamics: In a natural thermodynamic process, the sum of the entropies of the interacting thermodynamic systems increases.

- Third law of thermodynamics: The entropy of a system approaches a constant value as the temperature approaches absolute zero.

Adam Smith outlined three natural laws of economics:

Adam Smith	
	Three Natural Laws of Economics
	1. Law of Self-Interest - People work for their own good.
	2. Law of competition - Competition forces people to make a better product.
	3. Law of supply and demand - Enough goods would be produced at the lowest possible price to meet the demand in a market economy.

Figure 2 *Laws of Economics*—Three natural laws of economics by Adam Smith: Law of self interest; law of competition and law of supply and demand.

Laws of economics (Mimkes's statements)

1. First law of economics: Profit is a non-total differential form that depends on the path of acquisition.

In plain English: There are millions of ways to make money. Some may become rich overnight by winning lotteries, and some may toil their whole life and still be poor (see figure 3).

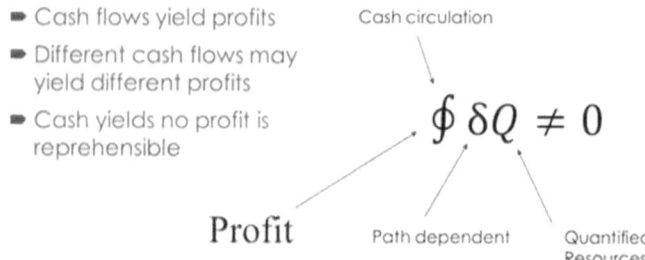

1st Law of Econophysics:
Money likens Living Water

- Cash flows yield profits
- Different cash flows may yield different profits
- Cash yields no profit is reprehensible

Cash circulation

$$\oint \delta Q \neq 0$$

Profit   Path dependent   Quantified Resources

Figure 3 *1st Law of Econophysics*—Mimkes postulates first law of econophysics with the notion that money likens a living water. Cash flow generate profits.

2. Second law of economics: The mean capital or standard of living is the integrating factor of profit and leads to the entropy of the capital distribution.

In plain English: A sense of self-worth depends not only on the amount of money to dispense at hand but also on the level of the living standard.

2nd Law of Econophysics:
Production Function (Entropy)
(Law of Competitions)

- Economic temperature, T, is an arbitrary value standard
- Production function or entropy is a measure of unavailable resources according to chosen value standard T
- Entropy is considered as the uncertainty or disorder of a system
- Entropy => degradation, e.g.: thieves, moths, rust, etc.

Production Function (Entropy)

$$dS \equiv \delta Q / T$$

Economic Temperature

$$\oint \delta Q = \oint T dS \neq 0$$

Figure 4 *2nd Law of Econophysics*—Mimkes postulates the second law of econophysics by introducing entropy as a production function and temperature a measurement of competition. It may be construed as the law of competition.

3. Third law of economics: Work increases capital and reduces capital distribution (work collects capital by distributing goods). Periodic work is always connected to two different economic levels.

In plain English: The third law of thermodynamics implies that in a barter economy, people of the same trade (e.g., salmon fishermen) do not trade with one another; therefore, $T = 0$ and $S = 0$ while the third law of economics states people get paid by working ($\Delta S > 0$) and income elevate living standard and self-worth ($\Delta T > 0$).

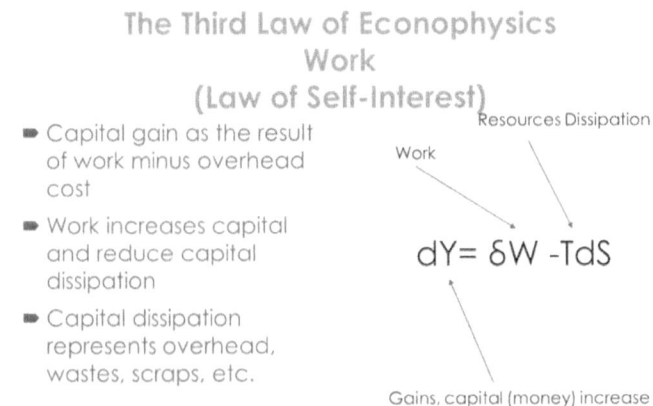

The Third Law of Econophysics
Work
(Law of Self-Interest)

- Capital gain as the result of work minus overhead cost
- Work increases capital and reduce capital dissipation
- Capital dissipation represents overhead, wastes, scraps, etc.

Resources Dissipation

Work

$$dY = \delta W - TdS$$

Gains, capital (money) increase

Figure 5 *3rd Law of Econophysics*—Mimkes postulates the third Law of econophysics relating gain with work and expenses (a form of dissipation).

Econophysics is a vain endeavor if it is studied as an abstract subject in isolation from people as Hanauer (2019) and Volti (2017) pointed out. Money has no value and cannot flow without disciplined people, societal laws, financial rules and regulations, and a high degree of trust. Money is economic lifeblood and *living water* only when spiritual laws are obeyed.

Excerpt of Professor Mimkes' correspondence:

Your manuscript has many interesting features and cites my articles correctly. And indeed, my approach is more theoretically. But you know that economics has the important feature of looking into the future at things like income that are not yet known (only ex post). They differ from things we know ex ante. Standard economic theories do not mark this difference in contrast to physics and mathematics by exact and not exact differential forms and their Riemann and Stokes integrals.

The first law of thermodynamics is written in exact (dE) and not exact (DeltaQ and DeltaW) differential forms. This difference is very important and makes thermodynamics such a general theory, which is very difficult to understand for people outside the field and leads to miraculous terms like entropy.

Entropy or probability is the mathematical answer to questions for the future:

No doctor will be able to tell you how old you will get, but statistics will, if you look at the death statistics. This will not be your real age, but will give you an idea how old you might become.

In the same way many economic questions may be answered, but they have to be marked as probabilistic. And mathematics helps to calculate unknown facts by finding an integrating factor that will lead to a term that may be calculated by integration.

This approach makes the theory very general and widely applicable and not narrow.

APPENDIX B

Beyond Capitalism and Communism

But you, Daniel, roll up and seal the words of the scroll until the time of the end. Many will go here and there to increase knowledge.

—Daniel 12:4

Since the first Nobel Prize in Economic Sciences was awarded to Ragnar Frisch and Jan Tinbergen in 1969, the Prize in Economic Sciences is awarded annually by the Royal Swedish Academy of Sciences, Stockholm, Sweden, according to the same principles as for the Nobel Prizes that have been awarded since 1901. There is a long list of outstanding scholars who contributed to the field of economic science. *However, the undeniable fact is that the world is yet struggling between the debate between capitalism and communism advocated by Adam Smith (1723–1790) and Karl Marx (1818–1883).* Prominent figures can only influence their contemporaries or future generations but not the people before them. Following is a chronological list of prominent figures who influenced the field of economics directly and indirectly. Notice how Adam Smith and Karl Marx are influenced by their contemporaries.

List of influential thinkers

1451–1506 Christopher Columbus
1473–1543 Nicolaus Copernicus
1483–1546 Martin Luther
1643–1727 Isaac Newton
1723–1790 Adam Smith (Father of Economics/capitalism)
1736–1819 James Watt
1759–1833 William Wilberforce
1769–1821 Napoleon Bonaparte
1789–1799 French Revolution
1809–1882 Charles Darwin
1818–1883 Karl Marx (communism)
1820–1895 Friedrich Engels
1833–1896 Alfred Bernhard Nobel
1839–1903 Josiah Willard Gibbs
1844–1906 Ludwig Eduard Boltzmann
1856–1922 Andrey (Andrei) A. Markov
1887–1961 Erwin Schrödinger
1879–1955 Albert Einstein
1882–1970 Max Born
1901–1976 Werner Heisenberg
1915–2009 Paul Anthony Samuelson
1917–2003 Ilya V. Prigogine
1928–2015 John Forbes Nash Jr.

Even to this date, economists and policymakers continue to debate over the thoughts of capitalism inspired by Adam Smith and communism initiated by Karl Marx. (See Table 1):

Table 1 *An Ongoing Debate*—It is a tragic irony that the debate of Capitalism and Communism/Socialism continue to dominate the political arena to this date regardless the development of economic theories in academia.

	Capitalism	**Socialism**
Ownership	Assets owned by private firms	Assets owned by government/ co-operatives
Equality	Income determined by market forces	Redistribution of income
Prices	Prices determined by supply and demand	Price controls
Efficiency	Market incentives encourage firms to cut costs	Government owned firms have fewer incentives to be efficienct
Taxes	Limited taxes/ limited government spending	High progressive taxes/ Higher spending on public services
Healthcare	Healthcare Left to free-market	Healthcare provided by government free at point of use
Problems	Inequality, market failure, monopoly	Inefficiency of state industry, less incentives
Advantages	Dynamic economy, incentives for innovation and economic growth	Promotion of equality. Attempt to overcome market failure

Neither capitalism nor communism is a perfect system. Just like turbine engines and piston engines, they are created for different purposes and environments. ***It is time to look beyond the dichotomy of capitalism and communism leaving capitalism and communism behind to focus on Divine Provisions: i.e., light, water, air, and all resources God blesses us. The following figure illustrates how communism and capitalism mapped out on a continuum of TS diagram along with some recent figures: Karl Marx, Sun Yat-sen, Max Weber, Andrew Carnegie, John Rockefeller, and Warren Buffett. They are figures who profoundly influenced our generation in Taiwan directly and indirectly.

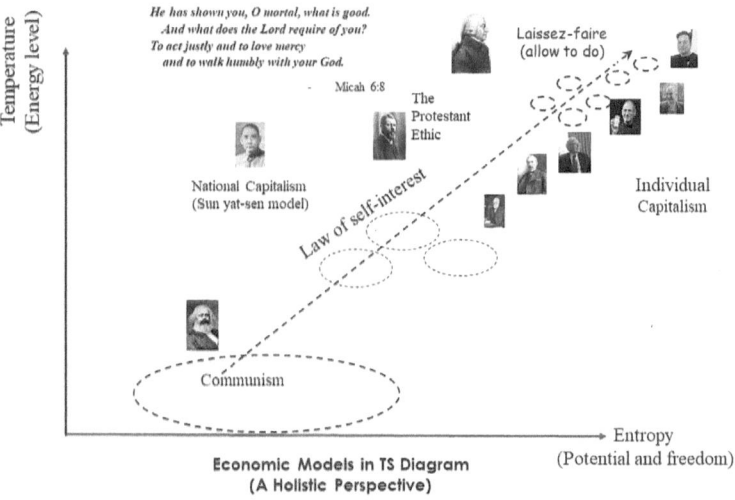

Economic Models in TS Diagram
(A Holistic Perspective)

Figure 1 *A Holistic Perspective*—Exemplary thinkers and practitioners in recent history mapped on a TS diagram schematically to illustrate the effect of the law of self-interest.

***Misinformation leads to misunderstanding, misunderstanding leads to and misconducts and misbehaviors. Recent shooting incident in Geneva Presbyterian Church outside Laguna Woods Village, California on May 16, 2022 sends an urgent message that it is time to stop the ideological debate of capitalism and communism and focus on God's plan. History and Heisenberg's uncertainty principle point out that communism is prone to economic crisis. On the other side, capitalism without Christian ethics leads to social injustice. Christian ethics is summarized in Micah 6:8b—Do justice, love mercy and walk humbly with God. Christian ethics can only be upheld by honoring the tripartite government outlined in Isaiah 33:22: "For the Lord is our judge, the Lord is our lawgiver, the Lord is our king; it is he who will save us."

Figure 1 correlates prosperity with the degree of freedom in government structures, providing a spectrum to chart economic theories and practices. Since the first Industrial Revolution, governments and economists have struggled under the long shadows of capitalism by Adam Smith and communism by Karl Marx. A closer examination of Keynesian and Samuelson neoclassical economics reveals the extended debate between the two schools. Figure 1 suggests that the choice of economic system, government structure, leadership, and lordship is crucial.

> ".....But as for me and my household, we will serve the Lord" (Joshua 24:15).

Recent bank crises, the Russia-Ukraine war, and the pandemic have exposed the vanity of human wisdom. "Truth" in Hebrew is "אמת" (emeth). If we take God (א, Aleph) out of truth, it becomes the Hebrew word "death" "מת" (mavet). Political-economic theories without considering God are doomed to fail. God's instructions are clear: "He has told you, O man, what is good; and what does the Lord require of you but to do justice, to love kindness, and to walk humbly with your God?" (Micah 6:8). Such Christian ethics can only be upheld by honoring the tripartite government outlined in Isaiah 33:22: "For the Lord is our judge, the Lord is our lawgiver, the Lord is our king; it is he who will save us."

APPENDIX C

Phenomenological Laws

And we know that in all things God works for the good of those who love him, who have been called according to his purpose.

—Romans 8:28

In physical science, transport phenomena of irreversible processes, such as heat flow, fluid flow, electrical flow, and frictional motions, can be expressed in a universal phenomenological relationship and described by linear laws, such as Fourier's law, Fick's law, and Ohm's law. These processes incur energy losses and consequently generate entropy (see figure 1).

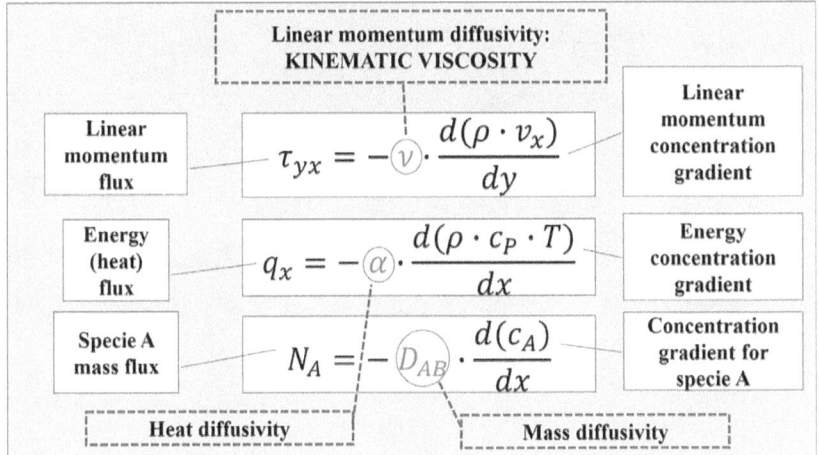

Figure 1 *Phenomenological Laws of Irreversible Processes*—Fick's Law of Diffusion, Fourier's Law of Heat Conduction, Newton's Law of Viscosity, and Ohm's Law of Current Flow are examples of phenomenological laws and have shared mathematical form.

In social science, economic development follows the trends of technology diffusion or diffusion of innovation. Though the mechanisms are vastly different, they share the same macroscopic characteristic as discussed in chapter 5.

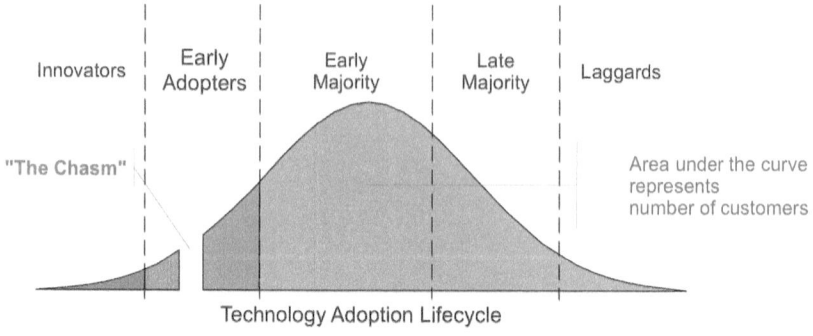

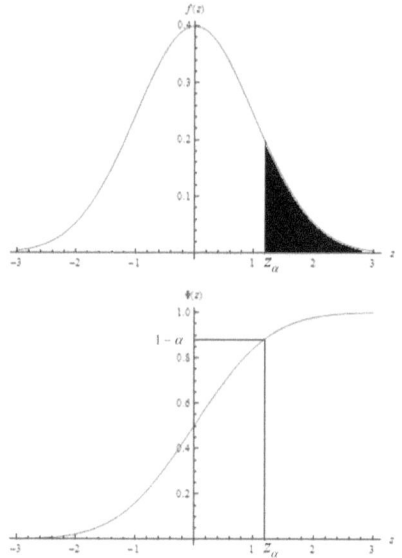

Figure 2 *Technology Adoption Lifecycle*—Economic development generally follows the trends of technology adoption lifecycle (top) which is a form of diffusion process can be described with a normal distribution. Bottom image showing the relationship between technology adoption lifecycle (normal distribution) and the growth curve (S-curve).

Phenomenological laws:

$$J = L X$$

{Rate of changes} = {coefficients matrix} × {driving forces}

where J is a vector with formed by $J_1, J_2, J_3, \ldots, J_n$, and X is a vector of $X_1, X_2, , X_3, \ldots, X_n$; and L is a n by n matrix with elements $L_{11}, L_{12}, \ldots, L_{nn}$.

Entropy production associated with these fluxes assumes the following form:

Entropy production rate = sum of {fluxes × forces}

Or

$$dS/dt = \sum X J$$

Metaphorical Phenomenological Laws

Metaphorically, phenomenological laws provide a convenient model for cash flow, goods flow, and information flow, etc. in economic systems. They are all irreversible processes in nature and produce entropy according to linear laws. For instance, cash transactions are subject to losses; goods in transit are subject to damages or losses; and information exchanges are subject to noises, miscommunication, misunderstanding, etc. (see figure 3).

Interestingly, the Scripture made it clear long ago in Matthew 6:19–20:

> Do not store up for yourselves treasures on earth, where moths and vermin destroy, and where thieves break in and steal. But store up for yourselves treasures in heaven, where moths and vermin do not destroy, and where thieves do not break in and steal.

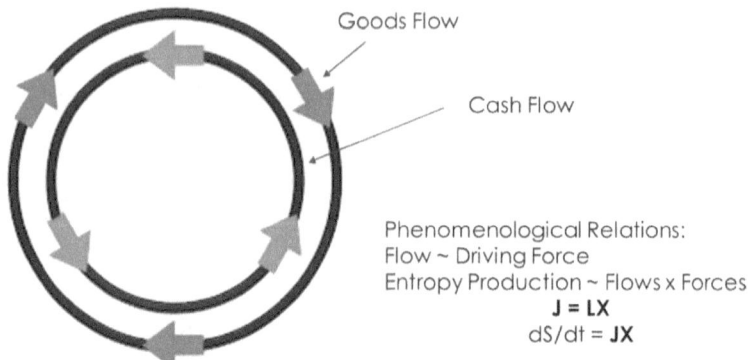

Figure 3 *Phenomenological Relations in Economics*—Note currency and goods/service flows are depicted moving in opposite directions. These two flows are correlated by an 'invisible hand'.

The "invisible hand" in Figure 3 can be modeled by the coupling coefficients L_{ij}. (Note: Lars Onsager proved that $L_{ij}=L_{ji}$ for thermoelectricity, electrokinetics, etc. and received the 1968 Nobel Prize in Chemistry.) However, what are the coupling phenomena in economics? Notably, relationships, value, credit, trust, risks (acts of God), cultures, information, etc., including laws, policies, rules, regulations, codes, advertising, brand names, reputation, trade secrets, intellectual properties, business intelligence, taxes, inflation rate, interest rates, etc. Bank crises, economic bubble bursts, depression, etc. are all caused by the breakdown of the coupling effects. There is much to be explored and researched to understand the "coupling coefficients" in the economic systems.

> *A change in motion is proportional to the motive force impressed and takes place along the straight line in which that force is impressed. (Isaac Newton Principia 1686)*

Table 1 summarizes some economic activities according to Newton's motion principle and phenomenological laws:

Table 1 Phenomenological laws examples

Flow	Driving Force	Governing Law
Goods	Surplus, shortage, technology diffusion, etc.	Fick's Law
Currency	Creative/innovative ideas, differentials in relationship, price, quality, laws, policies , taxes, interest rate, etc.	Fourier's Law
People	Migrant workers, outsourcing, living standard, acts of God, etc.	Newtonian's Law

APPENDIX D

The Art of Idealization

Imagination is more important than knowledge. Knowledge
is limited. *Imagination* encircles the world.

—Albert Einstein

Idealization is a valuable methodology in the history of science and
technology. Mathematicians have long adopted the strategies of aux-
iliary lines, imaginary numbers, and planes to aid problem-solving.
Newton related the fall of the apple to the moon in the sky via imagination. Later he postulated his first law: An object (the moon) will
remain in motion in a straight line unless acted on by an outside
unbalanced force (gravitational attraction). He deduced that the
force that pulls an apple to the ground also keeps the moon in orbit.
He stated: "Truth is ever to be found in the simplicity, and not in
the multiplicity and confusion of things." Issac Newton related the
moon and apple via imagination and postulated the laws of motion.

Figure 1 *Falling Apple*—Isaac Newton related the moon and apple via imagination and postulated the law of gravity and laws of motion. (Courtesy of Apple computer first logo)

Newton's thought process was widely adopted in the fields of science and engineering. Engineering students have been taught to use *free body* diagram and *control volume* to simplify complex systems via assumptions before numerical solutions and optimal design. (See figures 2 &3)

Free Body Diagram

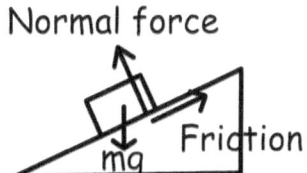

Figure 2 *Free Body Diagram*—Idealized 'Free Body Diagram' is a common approach to simplify complex problems via a set of assumptions in engineering problems.

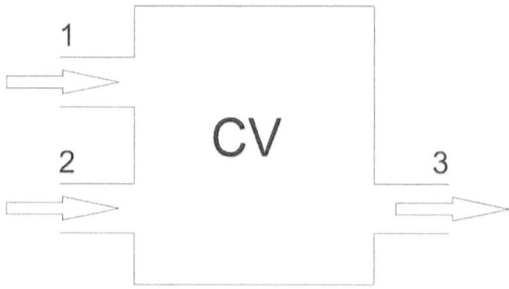

Figure 3 *Control Volume*—Choosing idealized control volume is the first step of problem solving and designing fluid systems, e.g.: pump, airplane, etc.

An ideal gas is an imaginary gas based on a set of assumptions of kinetic molecular theory (see Table 1). Note some assumptions for ideal gas are physically unrealistic (e.g., ideal gas particle has no volume, collisions between gas particles are elastic, and no kinetic energy is lost in elastic collisions). The role of ideal gas serves as a reference model to guide the understanding of the properties and behaviors of real gases (e.g., nitrogen, oxygen, and air).

Table 1 *Assumptions of Ideal Gas*—Comparison of an ideal gas and real gases. Note some assumptions for ideal gas are unrealistic.

What is an ideal gas? (Assumptions for ideal gas)	
Ideal gas	Real gas
No volume	Has volume
No molecular attractions	Has molecular attractions
Follow gas laws and at all pressures and temperatures	Differ at low temperature and high pressure

Likewise, mathematician and economist John Nash, who is known for game theory, proposed ideal money around early 1960. It paved the path for crytocurreencies, bitcoin, blockchain, etc.

Blessed are the pure in heart, for they will
see God. (Matthew 5:8)

REFERENCES

Bryant, J. 2011. *Thermoeconomics—A Thermodynamic Approach to Economics*. VOCAT International Ltd.

Carnot, S. 1824. *Réflexions sur la puissance motrice du feu et sur les machines propres à développer cette puissance* (in French). *Reflections on the Motive Power of Heat and on Machines Fitted to Develop That Power*. Paris: Bachelier.

Chakrabarti, B. K., Chakraborti, A., and Chatterjee, A. (eds). 2006. In *Econophysics and Sociophysics—Trends and Perspectives*. Wiley-VCH Verlag GmbH & Co.KGaA. Weinheim

Eftimov, T. 2013. Ideal Gas Laws in Thermoeconomics and Financial Bubble Formation. *Fundamental Sciences and Applications* vol. 19.

Encyclopedia of Materials: Science and Technology (Second Edition). 2001. Pages 3284–3287.

English Standard Version Bible. 2001. ESV Online. https://esv.literalword.com/

Faghri, A. and Zhang, Y. (Editors). 2006. "Thermodynamics of Multiphase Systems." *Transport Phenomena in Multiphase Systems*, pp. 106-176. Academic Press.

Georgescu-Roegen, N. 1971. *The Entropy Law and the Economic Process*. Harvard University Press.

Gibbs, J.W. 1873. "Graphical Methods in the Thermodynamics of Fluids" Part 1 and "A Method of Geometrical Representation of the Thermodynamic Properties of Substances by Means of Surfaces" Part 2. *Transactions of the Connecticut Academy* vol. 2, part 1, pp. 309–342 and part 2, pp. 382–404.

Hanauer N. 2019, October. The dirty secret of capitalism—and a new way forward [video]. TED Conferences. https://youtu.be/th3KE_H27bs

Harman, P. M. (Editor). 2002. *The Scientific Letters and Papers of James Clerk Maxwell*, vol. 3, pp. 1874–1879. Cambridge University Press.

Hubbard, W. H. J. H. Newtonian Law and Economics, Quantum Law and Economics, and the Search for a Theory of Relativity. Coase lecture, April 24, 2015. Law School Communications, The University of Chicago Law school.

J. W. Gibbs. On the Equilibrium of Heterogeneous Substances. Transactions of the Connecticut Academy of Arts and Sciences. 3, 108–248, 343–524, (1874–1878).

Kriz, R. D. 2008. Thermodynamic Case Study: Gibbs' Thermodynamic Graphical Method. http://sv.rkriz.net/classes/ESM4714/methods/Gibbs.html, accessed December 18, 2021.

Maslow, A. H. 1943. *A Theory of Human Motivation*. Psychological review. Vol. 50, no. 4, pp. 370–396.

Mimkes, J. 2006. A Thermodynamic Formulation of Economics. In *Econophysics and Sociophysics*, eds B. K. Chakrabarti, A. Chatterjee. Hoboken: Wiley

Prigogine, I. 1977. Time, Structure and Fluctuations. Nobel lecture. http://academic.pgcc.edu/~wboyle/TMDInfoEcon.pdf. https://www.youtube.com/watch?v=RrFSO62p0jk

Prigogine, I., Stengers, I. 1984. *Order Out of Chaos: Man's New Dialogue with Nature*. Bantam Books.

Richmond, P.; Mimkes, J.; and Hutzler, S. 2013. *Econophysics and Physical Economics*. Oxford.

Samuelson, P. Maximum Principles in Analytical Economics. Massachusetts Institute of Technology, Cambridge, Massachusetts Nobel Memorial Lecture, December 11, 1970.

Saslow, W. M. 1999. "An economic analogy to thermodynamics." *Am. J. Phys.* vol. 67, no. 12, pp. 1239–1247.

Shannon, C. E. July 1948. "A Mathematical Theory of Communication." *Bell System Technical Journal*. 27 (3): 379–

423. doi:10.1002/j.1538-7305.1948.tb01338.x. hdl:10338. dmlcz/101429.

Smith, A. 1776. *An Inquiry into the Nature and Causes of the Wealth of Nations*. London: W. W. Strahan and T. Cadell

Volti, R. 2017. *Society and Technological Change*. New York: Worth Publisher

Yen, M. S. 1975. *An Equation of State for Air For Temperatures from 80 K to 900 K and pressures to 2200 Atmospheres*. Master thesis. University of Idaho, Mechanical Engineering

Yen, M. S. 1979. A Study of Thermally-Induced Pumping of Carbon Dioxide Through Natural Rubber Membranes. Doctor thesis. Purdue University, Mechanical Engineering

Tribute

I have experienced several life-threatening crises in my life and saved by God's provisions and His Grace. I want to express my tributes to those people who prayed for me and took cares of me in those crises. Without their acts of love, I would not be here to endeavor this book. Specifically, I want to document the most relevant event:

At the end of 2019, My wife, Elaine, and I were on a Grand Princess cruise trip from San Francisco to Hawaii. On December 28th, I had life-threatening bleeding. I singed paperwork not to receive blood transfusion since I do not want to burden my wife and the family with undesirable consequences. On board doctor informed me the prospect of DOA (Dead on Arrival) since the cruise ship was in the middle of the route and no helicopter can reach us. The captain communicated with US Navy and National Coast Guard for a long range rescue mission. My salute to Grand Princess captain, medical team, US Navy, HSM-37 rescue crew, medical team in Hilo Medical Center, people who prayed for me in Fresno, San Diego, Hawaii, Taiwan, and China.

(https://www.navytimes.com/news/your-navy/2020/01/02/
navy-and-coast-guard-team-up-to-rescue-cruise-ship-passenger/)

The dramatic rescue operation in Pacific Ocean on January 1st, 2020 is the cue and a sign for me to endeavor this book. Prior to this incident, I have already abandoned the book project since I was discouraged to learn about the struggle and suicide of renowned physicist Ludwig Boltzmann who formulated entropy in terms of statistic mechanics, as well as life's storms and unanswered questions related to this book. Like the *Story of Jonah*, I was once prompted and inspired again to embark the book project in this plagued world. Therefore, my ultimate tribute goes to the Almighty God whose wonderful love and mighty deeds saved me:

> "He reached down from on high and took
> hold of me; he drew me out of deep waters."
>
> (Psalm 18:16)

ACKNOWLEDGMENTS

Numerous people inspired, encouraged and supported me to endeavor this book. Specifically, I want to acknowledge the following individuals:

Professor Mimkes' pioneer work on econophysics that stimulated my interest to endeavor in this field. Dr. Richard Stewart and Dr. Richard Jacobsen of University of Idaho. They guided me on the project correlating data for an equation of state (EOS) for the air. They also introduced me to the world of statistics, computer numerical methods, and transport phenomena. Dr. Touloukian and Dr. Schoenhals of Purdue University who led me to explore the irreversible processes. Dr. K. C. Lin of National Taiwan University who taught me thermodynamics.

Many friends served as sounding board to shape and form ideas of this book. I wish I could credit them one by one. I pray that God will grant the opportunities in future books. Finally, I want to thank Christian Faith Publishing staff helping me with editing, page design and cover design.

ABOUT THE AUTHOR

Majored in mechanical engineering, at National Taiwan University, Matthew always wondered how technology developed since the first Industrial Revolution (1760–1820). He noted the importance of human factors in these technology and economic development processes. In 1974, Matthew went to the University of Idaho for graduate study. His master thesis topic was "An Equation of State for Air." He enjoyed learning statistics, regression analysis, numerical methods, transport phenomena, etc.

Later he went to Purdue University for his PhD, studying non-equilibrium systems and irreversible processes. Afterward, Matthew took a job designing refrigeration systems and ice machines. He always marveled at the phase change of water from solid, liquid, and steam. Such marvelous phenomena inspired Matthew to search for the Creator. In 1983, he was baptized to become a Christian in Minnesota.

Late 1980's Matthew got interested in the applications of computer aided design and computer integrated manufacturing. He taught CAD and statistics in Manufacturing System Engineering of St. Thomas University. Meanwhile, he explored the fields of neural network and artificial intelligence with faculty in the University of Minnesota. 1989 Matthew joined California State University, Fresno to endeavor Computer Integrated Manufacturing Institute sponsored by IBM Corporation. Such appointment lead to a career of teaching and research. During his tenure, Matthew taught an assortment of courses: CAD, electronics, computer programming, manufacturing systems, database, networks, geographical information system (GIS), research methods, technology & society, etc. His research interest includes: automation system in food processing, neural network applications, GIS applications in agriculture and education, econophyics, etc.

In the year of 2010, Matthew came across the book Order Out of Chaos: Man's New Dialogue with Nature by Ilya Prigogine—a Nobel laureate. Matthew noted the universality of phenomenological laws of irreversible thermodynamics. By that time, Matthew, already a Christian, began to search the relationship of irreversible processes and the scripture. During the pandemic of 2020, the illuminating truth that the economic system can be summarized in two sets of equations began to be dawned on him: equations of state for equilibrium system and equations of changes for non-equilibrium systems. Therefore, he decided to elaborate the truth about economics from the perspective of divine provisions via history and scientific manifestation known as "scientia ancilla theologiae".

(Next page—original cover design)
Theme: Bridge Over Troubled Water

Steam *aka* living water powered locomotive and industrial revolution as well as ensuing struggles between capitalism and communism. This book serves as a bridge connecting old and new perspective of economics over the troubled water in the river of history, which is caused by unruly human greed and misleading theories. The artwork highlights key points in the book: the similitude of the law of supply and demand and the equation of state of gas; the conjugated pair of PV & TS for an ideal gas, the S-shaped equation of change in the Morning Star box; Maslow Hierarchy of Needs; Chinese character 熵 (entropy) and information & entropy relationship.

What is a gas model in economics? And what are its implications and ramifications? Why is it necessary to use pressure/volume and temperature/entropy conjugate in an equation of state for gas? What is entropy and its significance? Is entropy merely a 'grimy' word for disorder, losses, noise, etc.? How may we use entropy in economics? How is it related to 'work'? Can we tame the unruly human greed?.... I submit to search for answers in the Power of Living Water.

Living Water

A Holistic Perspective of Microeconomics and Macroeconomics

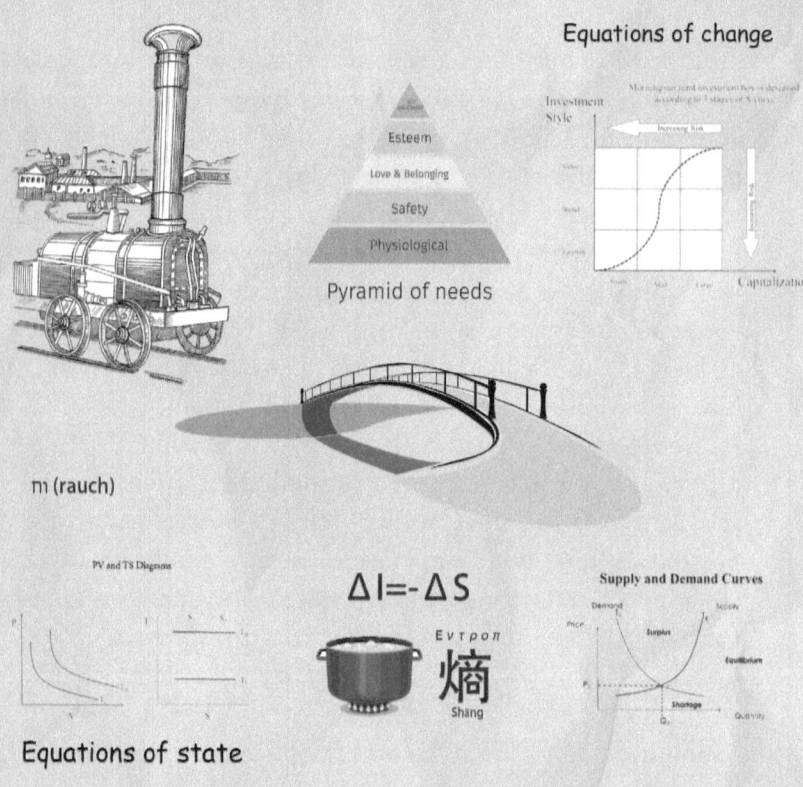

Equations of change

Pyramid of needs

ㄇ (rauch)

Equations of state

$\Delta I = -\Delta S$

Supply and Demand Curves

Matthew Yen